Y0-ABI-613

Isaac Newton
and Gravity

Sir Isaac Newton (1642–1727)

Pioneers of Science and Discovery

Isaac Newton
and Gravity

P. M. Rattansi, M.A., Ph.D.

Professor of the History and Philosophy of Science
University of London

PRIORY PRESS LIMITED

Other Books in this Series

Second impression 1978
SBN 85078 123 X
Copyright © 1974 by P. M. Rattansi
First published in 1974 by
Priory Press Ltd., 49 Lansdowne Place, Hove, East Sussex BN3 1HF
Text set in 12/14 pt. Photon Baskerville, printed by Photolithography,
and bound in Great Britain at The Pitman Press, Bath

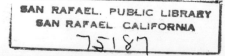
SAN RAFAEL. PUBLIC LIBRARY
SAN RAFAEL CALIFORNIA
75187

4

Contents

Illustrations

1. Newton's Early Years

Left During the plague of 1665, in which more than 100,000 people died, bodies had to be dumped into mass graves like this one at Holy Well Mount near London.

Below A London street scene during the Great Plague. Houses "visited" by the plague were marked by a cross, and charcoal, pitch and brimstone were burnt outside in the hope that this would kill it off.

By the autumn of 1665 the Great Plague of London was at its peak. Over 7,000 bodies were being thrown into mass graves every week. The plague began to spread into the eastern counties and struck the university town of Cambridge. Colleges were ordered to close and soon emptied of scholars and fellows. At night they were bathed in the glow of fires of charcoal, pitch and brimstone which were kept burning at some gatehouses. It was hoped that this would kill the "poison" of the plague.

Among those forced to seek refuge in the countryside was a young scholar of Trinity College, Isaac Newton. He went to his mother's farm at Woolsthorpe in Lincolnshire. Except for short visits to Cambridge, he spent nearly two years there. His teachers at school and university thought he showed great promise. But they could not have guessed what he was to discover in those two years, shut away most of the time in a lonely village. Newton made some of man's greatest scientific discoveries. His work was to give a firm basis to a new kind of science which replaced the ancient structure that had stood for two thousand years.

One of his discoveries was that white light was really made up of rays of different colours. Another was his method of "fluxions," or calculus as it is now known, which is a very powerful mathematical tool. The third, and perhaps the greatest, was to lead to the law of universal gravitation that ruled the fall of an apple to the earth as well as the motions of the planets around the sun. For a long time it seemed that Newton's greatest discovery would never be surpassed. The French physicist Pierre Laplace remarked in the late eighteenth century that Newton was the luckiest scientist who had ever lived, for there was only one system of scientific laws which explained the workings of the universe, and Newton had found it. After the scientific upheavals of the early twentieth century we no longer think so. But Newton still rules the way in which we analyse and predict the motion of physical objects in all kinds of circumstances.

Soon after Newton died in 1727 a famous French mathematician wondered if Newton could really have walked, eaten and slept like other men. What sort of man was Newton? What path led to his stupendous discoveries? How were they greeted by others? These are some of the questions we shall explore in this book. To grasp Isaac Newton's true greatness we must try to see what science was like before his discoveries

Top right Isaac Newton's birthplace at Woolsthorpe, a small village near Grantham in Lincolnshire.

Bottom right Isaac Newton at the age of twelve. This is from a painting made long after Newton's death by Frederick Newenham (1807–59).

changed its face. That is one of the hardest, but the most valuable, rewards of the time-travel which is history. Newton did not work on scientific problems alone, but spent a great deal of time studying Biblical prophecy and alchemy. Strange pursuits indeed for one of the founders of modern science! But for Newton, his scientific work was just one aspect of his search for an understanding of God's great plan as it was revealed in nature and in history.

Isaac Newton was born on Christmas Day, 1642, at the small manor of Woolsthorpe, about six miles south of Grantham in Lincolnshire. He was the son of Robert, a farmer, and his wife Harriet. Robert had died only a few months after his marriage and three months before Isaac's birth. Born prematurely, Isaac was said to have been so small that he could have been put into a quart mug; he was so weak that two women sent to fetch medicine for him did not expect to find him alive on their return.

Above The free school at Grantham, founded by King Henry VI, where Newton went as a boy.

When Isaac was three, his mother married the Reverend Barnabas Smith, a well-to-do rector, and went to live in the neighbouring village of North Witham. Isaac remained at Woolsthorpe and was brought up by his grandparents. His mother returned with the three children of the second marriage on the death of the Reverend Smith when Isaac was fourteen.

Newton attended two little day schools, and went to the King's School of Grantham when he was about twelve, lodging at the house of an apothecary. A seventeenth-century writer described him as a "sober, silent, and thinking lad," who was set apart from his school-fellows by his precocity and inventiveness. He made some remarkable mechanical toys and models, such as a windmill, a waterclock, and a carriage moved by the operation of a handle by the person who sat in it. He studied how paper kites could be made to fly higher and be more manoeuvrable. The paper lanterns he attached to them frightened country folk who thought they were comets. He marked the passage of the sun in the yard of his lodging and made sundials. Although time spent like this sometimes made him lose his place in class, he was clever enough to regain it as soon as he set his mind to it.

Above Inside the King's School, Grantham. This old building has, since Newton's time, been used as an assembly hall, gymnasium and store-room, and is now part of the school library.

Above Part of the window sill at the King's School where Isaac Newton carved his name.

When his mother returned to Woolsthorpe, Newton had to leave Grantham. As the eldest son, he was expected to manage the farm. But he soon convinced his mother that he would never make a good farmer. Sent to graze the sheep, he was often found beneath a hedge with a book, completely unaware of the sheep that wandered away. When he went into Grantham on market day to sell farm produce and buy things for the family, he would leave the job to the trusted servant who went with him and hide himself in the garret of his old lodging to read old books until the servant came back. Either his Grantham schoolmaster or a maternal uncle persuaded his mother, "What a loss it was to the world, as well as a vain attempt, to bury so extraordinary a talent in rustic business." So it was decided that Newton should go back to Grantham and prepare for university.

The years of Newton's childhood and early youth were stormy ones. They saw the outbreak of civil war, the beheading of Charles I and the rule first of the Commonwealth and then of the Protectorate of Oliver Cromwell. In his last month at Grantham in 1660, church bells proclaimed the restoration of monarchy and the triumphal return of Charles II to rule as King of England.

Newton's childhood saw the beheading of Charles I (*left*) in 1649, the rise of Oliver Cromwell (*centre*), and the restoration of Charles II (*above*) in 1660.

Newton's family are unlikely to have rejoiced when the Puritan party defeated the King in the civil war. Yet in later life Isaac continued to show many of the qualities which we associate with the Puritans: Biblicism, thrift, discipline, hard work, and an avoidance of noisy and uproarious company. These qualities must have made a deep impression on his character in those formative years. He was already eighteen when he entered university; many boys were enrolled at sixteen or even younger, and Isaac was much more set in his ways than they were.

He was admitted to Trinity College, Cambridge, on 5th June, 1661. When his mother married for the second time, she had insisted that her husband should give Isaac a certain sum of money, but it was not large enough for him to enrol at university except as a subsizar, which involved waiting upon his tutor at the dining table and doing a few other chores.

The Cambridge to which Newton came was busily adapting itself to the return of the Stuart monarchy.

Above Newton's notebook, begun while preparing for university in Cambridge. The right hand page shows his expenses on the journey from Woolsthorpe to Cambridge.

The two universities of Oxford and Cambridge trained the men who were at the centre of affairs in state and church. When the old order was overthrown about half the heads of colleges and fellows lost their posts because they would not swear an oath to the new Commonwealth. Now those put in their place feared similar treatment. Square mortarboards reappeared on the heads of dons and students in place of the round caps which had marked the Puritans ("Roundheads"). A Royalist academic joked that Puritan dons who hastily changed their caps had solved an ancient mathematical problem: they had "squared the circle."

We are lucky in the detail with which we can trace the development of Newton's ideas from the time he came to Cambridge. He seems to have kept almost every scrap of paper on which he ever wrote. It will be many years before historians have gone over all of the millions of words he left behind in his own hand. But the way his mind developed is reasonably clear.

The university statutes required him to learn his science from the works of the famous Greek thinker, Aristotle, who had lived in the fourth century B.C. Isaac's notebooks discuss the ideas of Aristotle. But they also discuss the ideas of more recent thinkers, like the Italian Galileo Galilei and the Frenchman René Descartes. These thinkers had challenged Aristotle and suggested that nature must be studied and explained in quite a new way. Newton studied the ancient ideas because he would be examined on them. He was not a slavish follower of the more recent thinkers and made some shrewd criticisms. Still, he took it for granted that scientific study must now follow the path they had set.

Let us now take a brief look at the scientific systems of Aristotle and of the more recent thinkers, to get some impression of the world of scientific ideas when Newton first came across it.

Left The famous nineteenth century cartoonist George Cruickshank's view of Newton. It shows Newton (who did not in fact smoke) using his lady friend's little finger to tamp down the tobacco in his clay pipe. The woman is possibly his reputed sweetheart, a Miss Storey of Grantham.

16

2. The Teachings of Aristotle

The science that Newton was taught at Cambridge University was based on the teachings of Aristotle. Some changes had been made in the thirteenth and again in the sixteenth centuries to make those teachings agree with Christian religious dogmas. Otherwise those ideas had remained much the same after nearly two thousand years.

How was that possible? It is easy to think of Aristotle as someone who had stopped scientific advance for two millennia. Joseph Glanvill, who lived at the same time as Newton, said that Aristotle's science had no support "either from sense or reason," and that it helped neither "knowledge or life." Why then had the Christian world accepted Aristotle since the thirteenth century as "the master of those that know," as the poet Dante Alighieri had called him? And why had he been so honoured even earlier by Jews and Moslems in the Middle Ages?

The mystery is even more puzzling when we remember that Aristotle's ideas clashed with some basic Christian teachings. Aristotle did not believe that the universe had been created out of nothing by God. His God did not watch over or intervene in the universe. Nor had he given men souls that survived the deaths of their bodies. Why, in spite of these differences, had Christian thinkers in the Middle Ages "baptized" Aristotle (as was mockingly said in the Renaissance)?

Francis Bacon, an English writer and philosopher, who had strongly opposed Aristotle's science in the early part of Newton's century, had tried to explain how Aristotle had come to rule men's minds. He said that when the barbarian peoples had destroyed the Roman Empire, human learning had suffered a

Below Francis Bacon (1561–1626), the English writer and philosopher and an opponent of Aristotle's theories of the universe. He was, like Newton, educated at Trinity College, Cambridge.

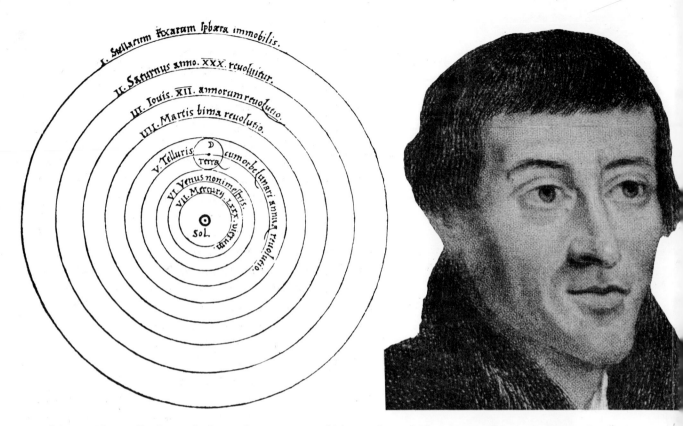

Left The picture of the universe put forward by the Polish astronomer Nicolas Copernicus (1473–1543) above. This arrangement had the earth and planets moving round the sun opposing Aristotle's view that the sun and planets moved round the earth.

shipwreck. All knowledge that was solid and worthwhile had sunk beneath the waves of time. Only that which was lightest and most worthless, like the systems of Aristotle and his master Plato, had floated to the surface like wooden planks.

Bacon's suggestion seems to ring true when we remember how alike the "New Science" of Galileo and Descartes was to the ideas of some ancient thinkers who had lived even before Aristotle. The "New Science" had three key features. It accepted the *cosmology* or arrangement of the universe put forward in the mid-sixteenth century by Nicolas Copernicus, in which the planets and the earth moved around the sun. It believed that the greatest secrets of nature would be won by revealing the mathematics that lay at their root. It supposed that all matter was made up of small invisible particles, and that the ways in which one piece of matter was different from another could be explained by the way in which those particles were arranged and moved. This is known as *atomism*.

All these ideas were found among various Greek thinkers who had lived before Aristotle, even if they were not brought together in the way in which they were in the "New Science." Aristarchus of Samos in the fourth century B.C. had denied what most people believed, and said that the earth and the planets move around the sun. Pythagoras in the sixth century B.C. had been so struck at how number governed music that he and his followers had made a religion out of mathematics. Atomism had been boldly used by Democritus in the fourth/fifth centuries B.C. to explain the birth of the universe and, indeed, everything that happened in nature.

Why had men turned their backs for centuries on what might seem very promising beginnings, and accepted instead the very different ideas of Aristotle?

The answer is surprising. Nor is it the one given by Bacon and others in the seventeenth century. It is that Aristotle's ideas were far closer to our commonsense picture of the world than any rival bodies of ideas. But Aristotle did not just dismiss these other ideas. Indeed, we know about some of them only because he discussed them in such detail. But in each case he gave reasons for believing that they were false. His reasons seemed very strong for a long time and we must try and see why.

Aristotle was, above all, a naturalist and biologist. He has a good claim to be called the founder of biology as a science. His study of living things made him believe that what was most striking and important about nature was the way in which everything seemed to aim at a particular target, and marched towards a goal. Every natural thing had a mature "form" and, when it was left free, moved towards it. The acorn became an oak, a baby became an adult man or woman. In studying living things, the most sensible question to ask about their parts was: what is it for? What function does it serve? Purpose and order marked nature in every way.

But that was quite contrary to the fantastic universe that the atomists had suggested, where chance ruled everything. Could atoms moving about in an unlimited empty space and becoming entangled and separated by chance alone really have led to the marvellous purpose and order displayed by nature?

Nor did Aristotle accept mathematics as a most important tool for studying nature. It is useful to know how long and broad, thick or thin a thing is. But such information is by no means the most useful when we want to understand what a thing is really like. Do we really know more about the snub-nose which the great Greek thinker Socrates had if we talk about it in terms of concave curves? Aristotle focussed his attention mainly on living and growing things. Mathematics seemed to be of little help in studying them. He never realized just how much mathematics could do for science. He insisted that we ought to study the world in all its living richness, rather than the pale ghosts of mathematics.

Aristotle rejected the strange idea of Aristarchus that it was the earth which really moved around the sun. Strong objections had already been quite reasonably made against it. We see the sun rising every day on the eastern horizon and setting in the west. And if there is one thing we can normally be quite sure of, it is that the ground is still and unmoving beneath our feet. There were also strong physical arguments against the rapid motion of the earth. All things not firmly attached to the earth would be thrown off into space, like water drops on the rim of a spinning wheel. Birds and clouds would move quickly westward as the earth left them behind. A ball dropped from a high tower would not fall vertically downwards because the earth would have moved while it was falling.

To these commonsense arguments Aristotle added

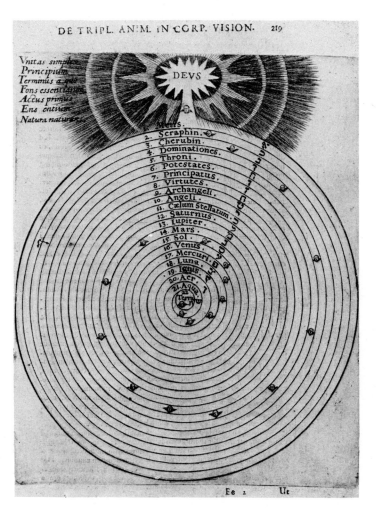

A corrupted view of Aristotle's universe (*see overpage*) mixed with Christian mythology. As with the Aristotelian view, the earth is at the centre surrounded by water, air and fire — but not in separate layers. Here the earth is at the centre of a great spiral which incorporates the planets, the stars, the nine orders of the angels, leading eventually to God.

others that were to carry great weight for a long time. Matter is observed on the earth as a solid, a liquid, a gas, or as radiant. Early Greek thinkers suggested that everything on earth was made of a mixture of four elements: *earth, water, air,* and *fire.* All four elements are present in any particular thing, but there is more of one than the others. Any one element can always be changed into another. Thus, the water we know is made of all four elements, but with an excess of the pure element *water*. When it is frozen, it becomes solid and is changed into *earth*. When heated, it turns into steam, which was taken to be *air*.

Aristotle made these earlier ideas more systematic. He noted that each element tended to move in a particular direction. A stone always fell down when released by the hand. Steam and fire always strove upwards. The purpose and order which had so impressed Aristotle when studying living and growing things seemed to him to be present in the motions of "dead" things too. As we have already seen, he believed that the true nature of a living thing was clear if we studied the mature "form" towards which it moved. The "elementary" nature of a thing, similarly, was shown by the direction in which it moved when not interfered with. Each thing moved towards its natural place in the universe. If the motion of the sun did not always mix up the elements, they would in time have settled down in their proper place. There would then be an earthy nucleus, surrounded by layers of water, air, and fire in that order.

It should now be clear that for Aristotle the earth *had* to be at the centre of the universe. That was the place of the heaviest and lowliest element. Newton was to ask why an apple fell down, rather than upwards or sideways. Aristotle would have answered that the apple's nature was earthy, and the centre of the universe was the natural place of earthy things.

The earth could not possibly move, according to Aristotle's ideas. Things moved because they were out of their place and wanted to get back to it. The earth was already in its proper place. "Natural" motions cease when things are in their places.

We see two sorts of motions on the earth. Either they are natural motions, and need no force. Or they are forced or "violent" motions, as when we throw a stone in the air, or blow down on a flame; or when a cart is pulled by a horse, or a boat rowed by oarsmen. In the second case, a push or a pull has to be supplied by a living thing which touches the thing being pushed or pulled. Both sorts of motion have a beginning and an end. Either a thing reaches its place, or

Aristotle's view of the universe, showing the layers of earth, water, air and fire. Much further out are the stars arranged in the signs of the zodiac and made up of the most perfect and superior of elements, the Ether.

that which pushes or pulls it stops doing so after a time.

But look up at the heavens. The sun and the seven planets revolve around the earth in unchanging circular paths. They are made up not of the four elements as is everything below the orbit of the moon, but of the most perfect and superior of elements, the *Ether*. It is right and fitting that they should move in circular paths, each with its own speed that is never too fast or too slow.

In this way, Aristotle split off the heavens from the earth. In the heavens, the home of things made of the most perfect element, there could be no change. There was only uniform circular motion, and motion in a circle has neither beginning nor end. It is like a changeless change. The earth, being made of an inferior element, could surely not take part in such a perfect movement. It was absurd to think that it could rotate around its axis to make night and day, and revolve around the sun to produce the changes of the seasons. These changes had already been explained by Greek astronomers as due to motion of the whole of the heavens around the earth once every twenty-four hours, combined with the sun's motion around the earth in the course of a year.

Aristotle's picture of the universe may seem strange to us because we have grown up with a very different one. But this short summary of his arguments should show that he brought together, arranged, and strengthened the conclusions which had seemed reasonable to men who were no less intelligent than ourselves. The sciences about which Aristotle wrote covered a great deal. They would include what we would think of as physics, chemistry, biology, geology, astronomy and cosmology. But he wrote on much more. He wrote books on logic, or the art of reasoning, morals, politics, and poetics which for a long time seemed to have said everything that man could possibly learn about in these fields. Almost everything that the Greeks, and their older neighbours like the ancient Assyrians and Egyptians, had learnt was brought together by Aristotle. And that was done not as an encyclopaedia, but in such a way that everything connects up with everything else.

It is not surprising that when the Christian world, cut off from ancient learning after the Romans left Britain in 450 A.D., prospered and attained a cultural level which made it appreciate these treasures of the mind which were now preserved by the educated Moorish conquerors of Spain, it could not ignore them. Like Jewish and Moslem thinkers before them, Christian thinkers, above all the Italian Christian philosopher St. Thomas Aquinas (c. 1225–74), were able to make changes in Aristotelian ideas sufficient for a time to prevent a headlong clash between them and basic Christian teachings. A complete higher education could be provided by the recovered works of Aristotle, together with commentaries which suitably changed them in certain ways to conform with current Christian doctrine.

3. The "New Science." Aristotle under Attack

Above René Descartes (1596–1650), the French mathematician, physicist and philosopher. His physical theories were firmly based on his philosophy, an approach to physics which Newton claimed to be quite contrary to his own which was based on the development of theories from experiments and experimental results alone.

Left Galileo Galilei (1564–1642), the famous Italian astronomer and physicist, receiving a visit in 1638 from the English poet John Milton (1608–74). Galileo was at this time under house arrest as his book, *Dialogo,* which favoured the Copernican over the Aristotelian system, had been judged by the Council of the Holy Office in Rome a challenge to authority.

While he studied ancient science to satisfy Cambridge University rules, Newton was at the same time reading with excitement the works of pioneers who barely twenty years earlier had proposed new ways of explaining nature. Two men who had blazed the trail were the Italian Galileo and the Frenchman Descartes.

What made their "New Science" so different from that of Aristotle? First of all, they insisted on the most careful testing of theories by observation and experiment before they were accepted as true. A good example is a famous experiment which was being eagerly discussed when Newton was an undergraduate.

Galileo had been struck by the fact that when a simple force pump was used to raise water from a well, the water never rose more than about thirty-four feet above the surface of the water. Soon after his death, his pupil Torricelli tried to explain why. Imagine that the earth is surrounded by a sea of air. The weight of the air must press down on all things submerged in this sea, just as the sea does in our oceans. The barrel of the pump is mostly empty of air when one end is in water and the plunger is raised. The weight of the air will press down on the surface of the water and force water up the barrel. The limit on the height to which the water could be raised must then be a measure of the pressure of the air.

Torricelli did not merely put forward a theory. He thought of ways of testing it. The pressure of the air supports a column of water thirty-four feet high. Mercury is about fourteen times as heavy as water. So the air should support a mercury column only 34/14 feet or 29 1/7 inches. Torricelli took a long glass tube and sealed it at one end. He filled it with mercury and turned it upside down in a bowl filled with mercury. The mercury fell until the column stood about thirty inches high. Torricelli had tested his theory and invented the first *barometer*.

Blaise Pascal, already famous in his teens as a mathematical genius, heard in France of the experiment. He thought of another test for the idea of a sea of air. A fish swimming upwards from the ocean floor will feel less and less pressure. The weight and pressure of the water increases with depth. So with the sea of air. The barometer should drop as we go up a mountain, since the air pressure steadily decreases. Pascal's brother-in-law made a careful experiment. A test barometer was left at the bottom of the mountain with an assistant. Another was carried to the top of the Puy-de-Dôme, a mountain in central France. It fell nearly three inches at the top of the mountain. The mercury column in the test barometer showed no change.

Torricelli and Pascal carefully tested out the theory of the sea of air. They chose to work with exact quantities to make the experimental tests as accurate as possible. That method was a vital part of the new scientific approach. We shall see later that Newton made the most careful and exact experiments to check his theories.

But it would be wrong to conclude that the modern thinkers believed in testing their ideas while Aristotle was an "armchair theorist." True, Aristotle did not set up exact experimental tests. But he believed in the detailed and painstaking study of nature, that is, primarily living things, the way a field naturalist does today. So careful and original were his observations, for example on certain sea creatures, that they were not improved on till the last century.

Experiments, after all, help to answer questions we put to nature. The sorts of questions we think worth asking, the answers we find satisfying, and the tests by which we make sure that our answers are correct:

Blaise Pascal (1623–62), the French mathematician, physicist and religious thinker who in the late 1640s and with the help of his brother-in-law proved experimentally the weight of air. Descartes claimed to have suggested the experiment to Pascal some years before.

these depend on certain ideas we begin by taking for granted. In the end, they depend on what we believe nature is like. If it is the tendency of things to aim at a goal, of living things to progress towards their mature "forms," that we take to be their most important feature, then (like Aristotle) we shall search, above all, for the "final cause" or purpose embodied in a thing. Once we have found it, our search is at an end. The test will be whether it agrees with observation, seems reasonable and satisfying to the mind, and whether we have arrived at it without making logical blunders.

The "New Science" made quite different assumptions. Descartes said nature was an automaton. It had to be studied not in terms of the living and growing organism, but as if it were a machine. Nature was matter in motion. That matter had certain mathematical features: size, shape, arrangement, and motion. Matter can affect other matter only by colliding with it. A scientific explanation should aim at giving us a mathematical account of nature in terms of matter in motion.

For Descartes, plants, animals, and human bodies were really marvellously constructed machines. The earth and the planets that whirled unceasingly around the sun were parts of a heavenly clockwork. When we study machines, it makes no sense to ask what purpose they are trying to fulfill. Machines can have none except those which their makers have built into them. One piece of matter cannot have greater dignity than another, as the stars and planets had over all others for Aristotle. Nor can circular motion be superior to motion in a straight line.

Descartes tried to imagine how the world could have been born by the operation merely of the two principles of matter and motion. In the beginning God could have created matter and divided it into large chunks which were then left to rub against each other. After that, as long as He kept the quantity of motion in the universe the same at all times, so that the machine of the universe never ran down, everything would take place purely mechanically. The rubbing of matter against other matter would in time produce three basic kinds of matter. The finest would collect and form the bodies of the sun and the stars, and swirl in immense *vortices* or whirlpools about them. Our sky would be made up of the second kind of rougher spherical particles. Finally, the earth, the planets, and the comets would be made up of the roughest, irregularly shaped, and slower-moving particles.

Descartes tried to explain the movements of the earth, planets and comets as due to the motions of the vortices. The same with gravitation on earth. Descartes' universe was a full universe, for wherever there was space there was matter. The vortex of particles of the second type of matter would cause a stone thrown upwards to sink again to the earth. Similar explanations were given by Descartes of such mysterious and baffling phenomena as the tides on earth, magnetism, and the beating of the heart in the animal body.

Few of Descartes' detailed and ingenious explanations are accepted today. Soon after his death it was widely pointed out that his own work fell sadly short of the new model of doing science which he had proposed in place of the ancient one. Descartes had concluded that the most certain kind of human knowledge was that of mathematics. If scientific knowledge of nature was ever to become more than a matter of opinion and guesswork, it must be reduced to mathematics. That was made possible by the mechanical view of nature. Matter could affect other matter only by colliding with it, and all that happened in nature was due to matter in motion. Armed with a set of mathematical laws to tell us what would take place when one piece of matter collided with another—the "laws of motion and of impact"—we should be able to explain and predict everything that happened in nature.

Descartes' universe which was developed in order to explain gravity and the motion of the planets. He pictured the universe as filled with whirling vortices of subtle matter with a star at the centre of each vortex. Gravity is caused, he said, by the pressure of this whirling matter towards the centre of each vortex.

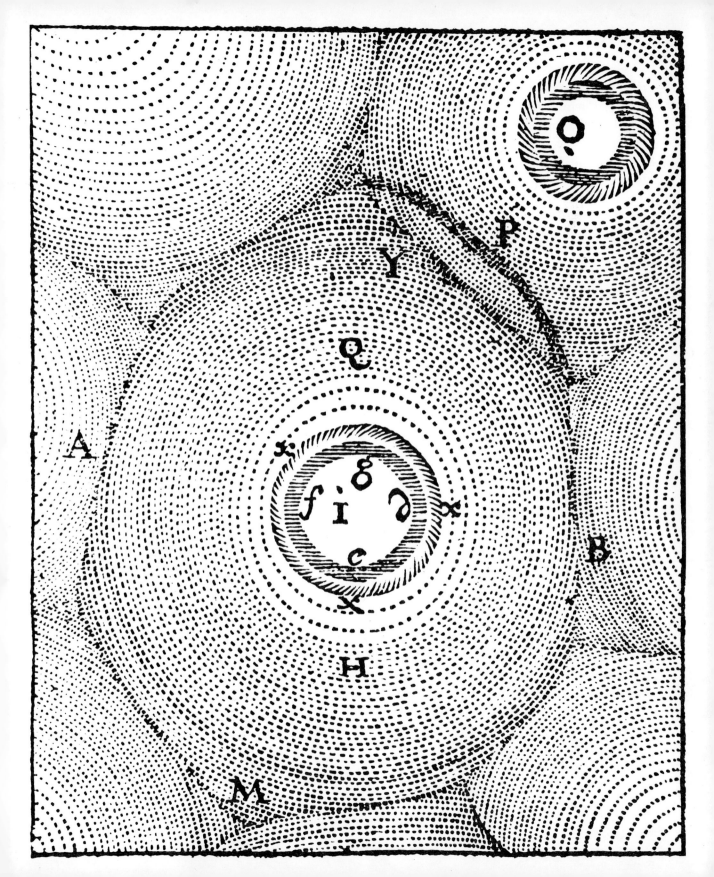

Descartes gave homely and easily pictured analogies which may have helped to make his ideas popular. He compared the motions of the planets to that of bits of cork caught up in a whirlpool, the reflection of light to tennis balls bouncing on a hard surface, and the action of the heart to the generation of heat in hay-mows. He did not work out his ideas mathematically as his own ideal of science demanded. Descartes had recognized that even if God had used only the two principles of matter and motion in running the universe, there were countless ways in which they could have been used to cause any particular phenomenon that science studied. Only experimental tests could tell us what mechanism lay behind any particular effect. But few of his own bold explanations had been put to searching experimental tests.

Despite these glaring faults that others were quick to notice, Descartes seemed to be a great liberator of the human mind to those who fell under the spell of his work. They included many very gifted young people. Why did Descartes make such a great impression upon their minds?

The "mechanical-mathematical" view of nature, which Descartes had spelt out so clearly and in so much detail, did not spring suddenly and fully-formed to replace Aristotle. The subtle mixture of Aristotle and Christian doctrines by medieval churchmen had already begun to fall apart in the fourteenth and fifteenth centuries. During the Renaissance men eagerly studied the writings of ancient Greece and Rome for models of thought and conduct. They showed contempt for the technical systems devised by churchmen with their eyes fixed wholly on the life to come. By the sixteenth century the Aristotle whose ideas lay behind these systems was being openly attacked. His was not going to be the last word on human knowledge.

It was one thing to attack Aristotle's system as out-dated, but quite another to replace it with a new one

Nicolas Copernicus (1473–1543), the astronomer who put forward the idea of a sun-centred universe. This woodcut, one of three made during his lifetime, shows him holding a Lily-of-the-Valley which normally signifies humility or purity.

with the same range and unity which would also be suitable as a new basis for higher education. There were endless arguments about the alternatives.

Two products of the turmoil in the world of ideas are of especial interest. John Donne, the English poet, imagined in 1611 some of the men who had overthrown long-established opinions putting their claims before the Devil in Hell. The two men of science he chose were the Polish Copernicus and the German Paracelsus. Copernicus claimed to have overturned "the whole frame of the world, and thereby ... [to be] almost a new Creator." Paracelsus had brought the whole of existing medicine into contempt, and had tried "not only to make a man ... but also to preserve him immortal."

Copernicus was the most outstanding astronomer of his time. In his masterpiece, whose first printed copy he is said to have received on his deathbed in 1543, he tried to simplify astronomy with the radical idea of having the earth and the planets circling the sun. This, he said, explained all changes seen in the heavens; he reworked the calculations by which astronomers predicted the motions of sun, moon, and planets. But he had no real answer for the physical objections to a moving earth (as we noted earlier) already put forward in ancient times. Copernicus could be accepted only when a new physics, compatible with a rotating and revolving earth, had been created to take the place of that of Aristotle.

Above Philippus Aureolus Theophrastus Bombastus ab Hohenheim (1493–1541), known more commonly as Paracelsus, the boastful Swiss alchemist, one of the founders of medical chemistry.

Above "The Stone of Saturn" from Michael Maier's *Atalanta Fugiens* (1618). This engraving, one of fifty which links Greek mythology with alchemy, shows Kronos (Saturn) vomiting up the stone which he had earlier swallowed thinking it was the baby Zeus (Jupiter). The stone was, of course, the very Philosopher's Stone sought by the alchemists. Newton spent a great deal of time and thought studying Maier's writings.

Quite different was the reform attempted by Paracelsus. For centuries alchemists had tried to turn base metals into gold. They claimed the support of Aristotle's ideas. If every thing developed towards its perfect form, then the metals "growing" in the earth must be ripening towards the form of gold. The alchemist wished to speed up that process by using a ferment, like the yeast used in bread-making and brewing. That ferment, called the "Philosopher's Stone," later came to be regarded as a "Universal Elixir." Just as it rid metals of their imperfections, so it would rid men of all diseases. Paracelsus' achievement was to turn alchemy away from the search for transmutation to chemical remedies for diseases. He

Right "The Alchemist," after a sixteenth century engraving by Jean de Vries. The alchemist himself is kneeling before an altar where there are some old alchemical or astrological books. The shelves and mantelpiece are lined with flasks and alembics and the musical instruments on the table illustrate the importance of music to alchemy.

tried to read all nature in terms of the reactions that boiled and bubbled in the flasks and alembics (early forms of still) of the alchemist's laboratory.

Paracelsus insisted that knowledge of nature must come from direct experience, not from the writing of Aristotle. That knowledge was to be used for human welfare. These ideas have a modern ring. Aristotle and the schoolmen had placed too much confidence in the power of human reason alone to penetrate the mysteries of nature. They had ignored the careful checks and experiments on which the "New Science" was to insist. Knowledge to them was for gratifying man's wonder and curiosity, rather than to furnish him useful inventions and discoveries. But the loss of confidence in Aristotle during the Renaissance often went with a revived belief in the "magic" of the ancient world. The stars were thought to govern our world; man had the power to tap their power and work wonders by a knowledge deeper than that

Wadham College, Oxford, where John Wilkins, Cromwell's brother-in-law, brought together a scientific group after his appointment as Warden in 1648. Its activities led to the founding in 1662 of a society given a royal Charter by Charles II. It was "The President, Council and Fellowship of the Royal Society for the Promotion of Natural Knowledge"—the "Royal Society" for short.

Two of the most outstanding founding Fellows of the Royal Society were Robert Boyle (*above*) and Sir Christopher Wren (*above right*). Boyle (1627–91) was something of an all-rounder with contributions in physics, chemistry and medicine. He is best known for his work on the properties of air. He differed from Newton in his views on light and colour and also on alchemy. Wren (1632–1723) was an architect and astronomer, famous for his design of St. Paul's Cathedral.

brought by using mere reason. The ideas of Paracelsus were tied to these magical beliefs.

There were many voices, calling on men to follow different paths, if the old beaten road of Aristotle was to be abandoned. When the old order in church and state was overthrown in England between 1640 and 1660, some of those who wished to reform education and society suggested replacing Aristotelian science by a Paracelsian one at the English universities.

That helps to explain the wide appeal of Descartes. At one stroke he seemed to have cut through the tangled forest of Aristotelian-scholastic ideas, while avoiding the mystical twilight by which many sought a new way to true knowledge of nature. His mechanical account of all that Aristotle had explained in nature gave others confidence in attempting a different account, even if his own explanations had to be improved or thrown out.

In England the new programme captured the

enthusiasm of a group of brilliant young men. They included Robert Boyle, a younger son of the Earl of Cork; his assistant, Robert Hooke, who was later to quarrel bitterly with Newton; and Christopher Wren, who advised on the rebuilding of London after the Great Fire of 1666 and designed St. Paul's Cathedral. At the Restoration these men founded the Royal Society for the Promotion of Natural Knowledge, which adopted the way of explaining nature mechanically. "New Science" seemed to have become respectable and had caught the interest and patronage of King Charles II and fashionable society.

Gresham College which was the meeting place of the Royal Society until 1710 when it moved into its own rooms in Crane Court, Fleet Street (see illustration on page 81). Sir Christopher Wren was Professor of Astronomy there from 1657 to 1661.

4. *Light and Colour.*
Newton's Début on the Scientific Stage

Like many other gifted young men, Newton had felt the thrill of seeing the world in the new light of Descartes' work. He was eager to try his powers in the exciting adventure that science seemed to offer. At the same time, he well knew Descartes' shortcomings. His ideas did not have the mathematical form his own ideal demanded. Nor had they been put to rigorous experimental test. In his undergraduate notes Newton sketched perpetual motion machines that drew their power from Descartes' whirling subtle matter (described on pages 27 and 28). But he also noted that Descartes' explanations of tides on earth by the action of that matter did not agree with the way they actually occurred.

Soon after he graduated, Newton tackled the basic problem that had to be solved before the new physics of matter and motion could be more than a dream. The mechanical programme had explained all changes in nature as changes in the position and motion of material bodies. Some mathematical way had to be found of pinning down that process of change. The crucial problem was to measure how quickly one quantity in a situation changes in relation to another quantity. *Velocity,* for example, is the rate of change of *distance* with respect to *time. Acceleration* is the rate of change of that rate of change, or how quickly velocity increases. Newton's mathematical genius took wings, he laid the basis of what we now call the *differential* and *integral calculus,* and so gave physics its most powerful mathematical tool. By *differentiation* it was possible to measure how quickly one changing quantity changed in relation to another at any given point in a process. *Integration* was the reverse process, just as subtraction is the reverse of

Left Trinity College: Nevile's Court and the Library. The Library was added in 1676 with funds raised by Isaac Barrow (*right*), the Lucasian Professor in Mathematics at Cambridge.

addition. If we know the rate of change, we can find by integration the changing quantities involved.

Newton's experimental cunning and precision were to be shown in his paper on light and colours that marked his sensational entrance upon the stage of science. All these qualities were to meet in his supreme discovery, the law of universal gravitation. This was to realize the dream of a physics compatible with Copernicus that had inspired men like Galileo and Descartes.

The roots of these epochal discoveries lay in the two year period (1665–66) that included the Great Plague and the Great Fire of London. Newton seems to have owed little to his formal education or teachers. He was almost entirely self-taught. University teaching was lax. Too many of the fellows were appointed as a reward for royal services, or were waiting for better jobs as clergymen elsewhere. Newton himself became a fellow soon after returning to a Cambridge freed of the plague. His achievements won the admiration of a senior member of his college, Isaac Barrow, the

Newton's own drawing of the reflecting
telescope of 1668. He presented the Royal
Society with a similar one in 1671.

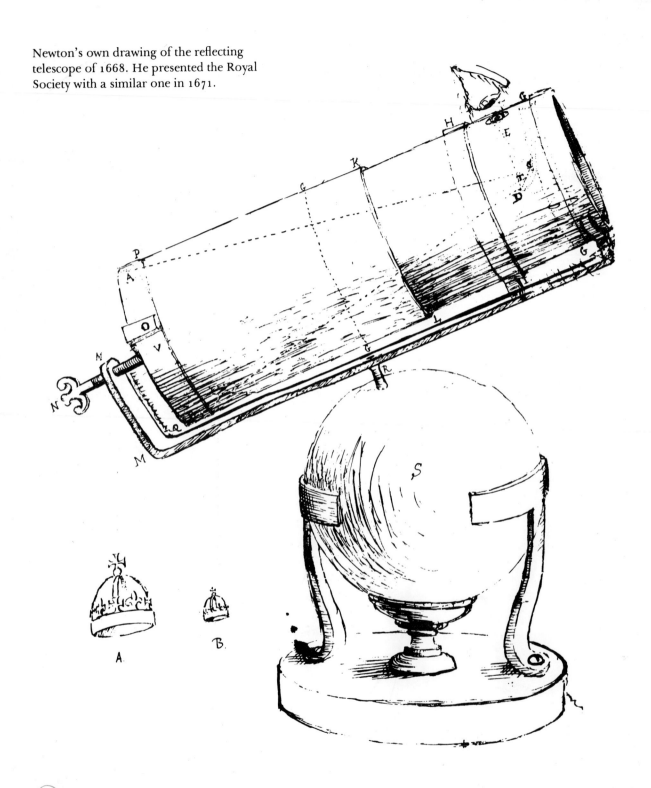

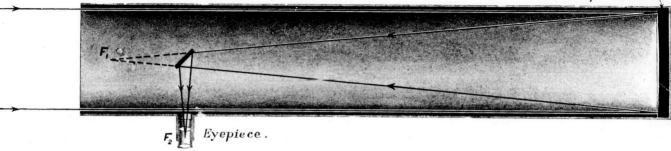

Concave speculum.

F_1

Eyepiece.

F_2

Diagram of Newton's reflecting telescope. This instrument almost totally eliminated chromatic aberration as the law of reflection is the same for all colours. A very small amount of splitting of colours occurred at the eyepiece.

Lucasian Professor in Mathematics. Barrow asked Newton's help in revising his own optical lectures for printing, and praised him in letters to his London friends. Newton was chosen soon afterwards to succeed him in the Lucasian Chair in Mathematics. The fame of a reflecting telescope made by Newton reached the Royal Society. Newton sent them the telescope as a gift. It created some excitement. Charles II is said to have looked through it. Newton was soon elected a Fellow of the Society.

Newton was pleased by the Society's appreciation. In a letter to the Secretary, Henry Oldenburg, he revealed that he had a far richer gift to offer than a telescope: a whole new theory of light and colours which had led him to make the instrument. Newton knew the importance of his discovery. He called it "the oddest, if not the most considerable detection, which hath hitherto been made in the operations of Nature."

The short paper Newton sent some weeks later is one of the most famous documents in the history of science. The opening passage conveys its flavour: "To perform my late promise to you, I shall without further ceremony acquaint you, that in the beginning of the Year 1666 (at which time I applyed myself to the grinding of Optick glasses of other figures than *Spherical*), I procured me a Triangular glass-prism, to try therewith the celebrated Phaenomena of Colours.

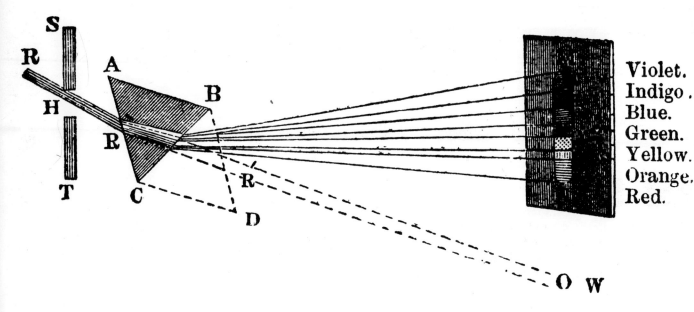

Violet.
Indigo.
Blue.
Green.
Yellow.
Orange.
Red.

And in order thereto having darkened my chamber, and made a small hole in my window-shuts, to let in a convenient quantity of the sun's light, I placed my Prism at its entrance, that it might be refracted to the opposite wall. It was at first a very pleasing divertisement, to view the vivid and intense colours produced thereby; but after a while applying myself to consider them more circumspectly, I became very surprised . . ."

The threads lead from this account, once more, back to Descartes. In his first published work Descartes had taken the problems of light as a challenge to and test of his new method. Even in ancient times optics had been found to be a science well suited to mathematical treatment. Moreover, ancient and medieval writers had found it useful when working out problems to think of light rays as if they were balls hitting a surface. These features fitted perfectly into Descartes' programme. He derived the *sine law of refraction* by imagining light rays as tennis balls striking a surface. He explained the formation of the rainbow.

Left Newton's prism experiment showing the splitting up of white light into its component colours.

Right Newton in his room investigating light.

42

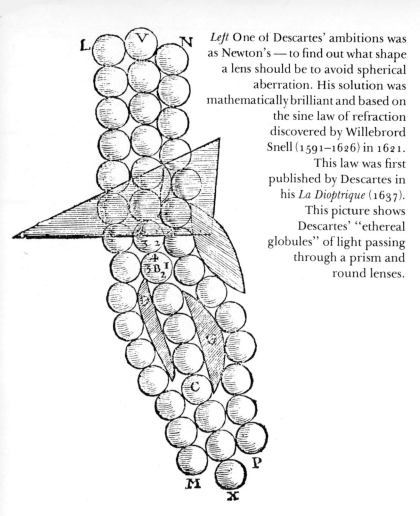

Left One of Descartes' ambitions was as Newton's — to find out what shape a lens should be to avoid spherical aberration. His solution was mathematically brilliant and based on the sine law of refraction discovered by Willebrord Snell (1591–1626) in 1621. This law was first published by Descartes in his *La Dioptrique* (1637). This picture shows Descartes' "ethereal globules" of light passing through a prism and round lenses.

Descartes did not believe that light was really made up of tiny solid particles. He thought of it as the pressure through the ether of the luminous matter which made up the bodies of the sun and the stars. The ordinary laws of mechanics applied to that pressure. How were colours to be explained? The "New Science" assumed that sensations like colour, taste, smell, and touch were due to the more basic mathematical features of matter. Descartes said that we see colours through the action on our optic nerve of "ethereal globules" which are set spinning like tops when they obliquely strike an object. Different speeds of rotation give rise to different colours.

One important reason for the new interest in optics was the improvement of telescopes. Galileo made one

Right Galileo's telescope. Galileo was, in 1609, the first to apply the newly invented telescope to astronomical observation. With it he revealed, amongst other things, mountains on the moon, many stars invisible to the naked eye and four of Jupiter's moons.

Left A lens-grinding machine of 1671. Though Descartes had come up with the solution to the problem of spherical aberration, grinding machines of the seventeenth century were still far too crude to grind accurately any shape with the non-spherical surfaces that he required.

in about 1609 after hearing reports of its invention. It consisted of an object lens, to form the image of a distant object, and an eyepiece to magnify the image. Galileo pointed his instrument at the heavens and made several surprising discoveries which he used to support Copernicus. Descartes had looked at the difficulties in improving the telescope. The chief one arose from what would now be called "spherical aberration." The lens gave a fuzzy image because light rays from the top and bottom ends of the lens did not come to a sharp focus with the other rays. Descartes suggested that the solution was to grind lenses of non-spherical shapes. That was why Newton in 1666 was trying what proved the near impossible task of grinding lenses "of other figures than spherical."

When Newton tried out the prism experiment, what first surprised him was the shape of the spectrum on the wall. It was oblong, and about five times as long as it was broad. He had expected to see the round image of the sun. He made careful experiments to find out why.

He passed the beam through various parts of the prism to see if the thickness of the glass affected the result. The shape stayed the same. To make sure the beam had not been scattered by a flaw in the glass, he placed another prism upside down behind the first one. By doing this any changes in the light beam due to its triangular shape would be cancelled out. However, if the image had been altered because of a flaw in the glass the changes would be even greater. Newton would have seen a longer oblong and perhaps a greater splitting up of colours. In fact, the image became round again, as if the beam had not passed through a prism at all.

Could the shape be due to rays from different parts of the disc of the sun meeting the prism from different angles? Precise calculations showed that this could not possibly have led to such a large angle between the

Experiment to show that the colours that appear when light passes through a prism do not split up any further when viewed through a second prism. From an engraving made in 1747.

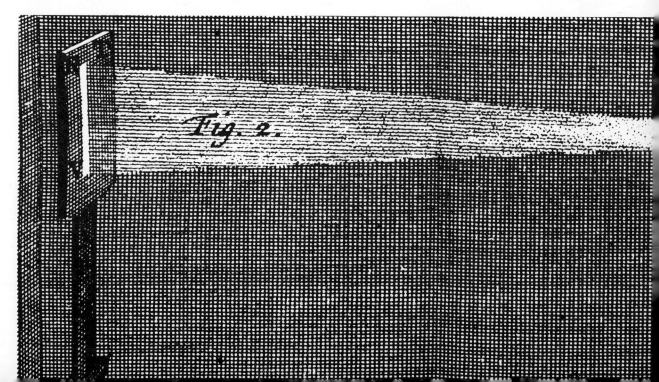

emerging rays. Could it be that the rays moved not in a straight but in a curved line when they emerged from the prism? Newton remembered (and there is a strong link here with Descartes' ideas) that he had "often seen a Tennis-ball, struck with an oblique racket, describe such a curved line." By letting the image fall on a board at various distances from the prism, he showed that light rays must travel in straight lines.

Having rejected these possibilities experimentally, Newton now tried what he called his "crucial experiment." He fixed a board with a small opening behind his prism. About twelve feet away he fixed another board with a small hole, and placed a second prism behind it. By letting in a beam of light, and turning the prism over he could select a ray of a single colour from the band of colours on the second board to pass through the second prism. Newton found that a ray of any one colour was unchanged by the second prism. But the lower colours of the first image were bent (refracted) more than the higher ones by the second prism.

This led Newton to a revolutionary conclusion,

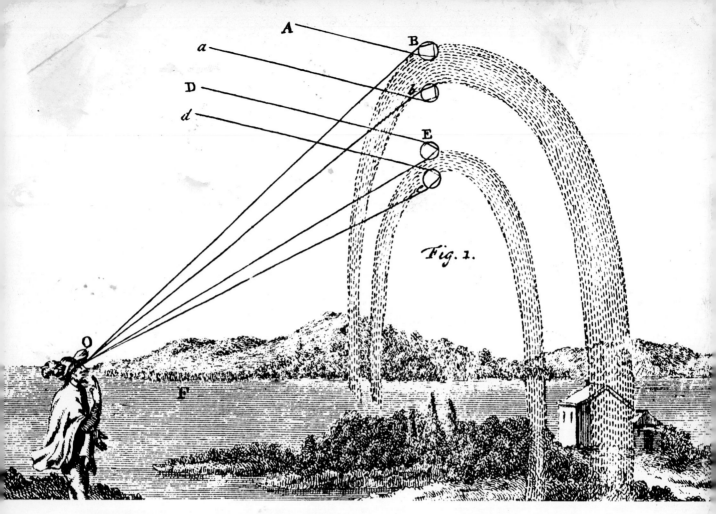

Fig. 1.

quite the opposite of what the ancient philosophers as well as his fellow men of science like Barrow believed about colours: "The true cause of length of that Image was detected to be no other, than that *Light* consists of *Rays differently refrangible* which . . . were, according to their degrees of refrangibility, transmitted towards divers parts of the wall."

Almost all existing theories of colours held that these colours were due to a mixture of light and shadow. Newton said no; white light itself is a *mixture* of colours. The prism does not produce colours by mixing white light with different amounts of shadow. It merely separates out the colours. Things look coloured because they "reflect one sort of light in greater plenty than another."

Although Descartes had made a valiant attempt to explain the formation of rainbows it was left to Newton to provide a full explanation in terms of the dependence of refraction on colour. This picture is from a "popular" version of Newton's theories which appeared in 1747.

William Blake (1757–1827) painted Newton (*above*) in 1795 and almost certainly in connection with his *First Book of Urizen* (1794). In it he describes a tyrannical kingdom ruled by the "materialist" philosophy of Isaac Newton, Francis Bacon and John Locke — a philosophy that he, together with many writers of the Romantic Movement, detested (see page 86). Blake believed that life should be based on "spiritual" laws and not on the "natural" laws of Newton. He described Newton's world as one of "soul-shuddering vacuum."

Newton's conclusions led him to abandon his "glassworks." He now believed that the main drawback of telescopes was what we would now call "chromatic aberration." Because white light is a mixture of "differently refrangible" rays, those rays could never be brought into a single focus by an object lens. Such an image would always have fuzzy coloured outlines. Reflection, however, does not have this problem. At this point Newton was forced to drop the subject. The plague came and he had to leave Cambridge. It would be two years before he was able to pick up the threads and set about trying to solve the problems involved in building a reflecting telescope using a concave mirror.

Newton's account of his astonishing discovery

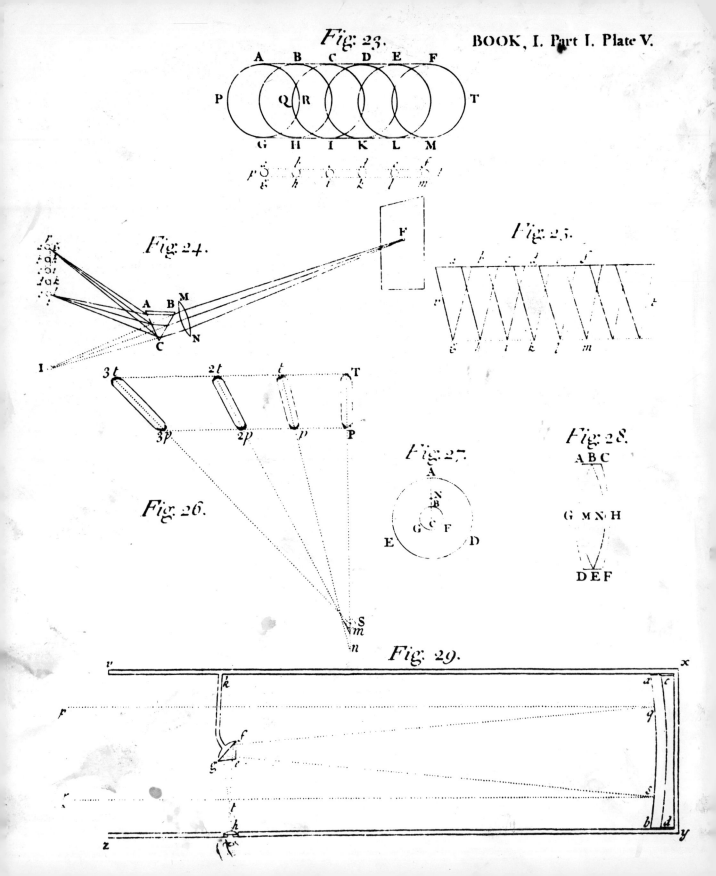

Fig: 23.

Fig: 24.

Fig: 25.

Fig: 26.

Fig: 27.

Fig: 28.

Fig: 29.

Right A bust of Newton in the Royal Observatory, Greenwich.

Left A plate from Newton's *Opticks* (1704).

seems a perfect example of the true method of science. Assuming no theory in advance, Newton had noted what took place when a beam of sunlight passed through a prism. His observations raised certain questions. To test various possible answers, he set up a number of careful experiments. Finally, a crucial test led to a fundamental conclusion.

But Newton's account is really a cleverly arranged argument for his theory rather than an historical account of its discovery. He had thought and worked on the problem for at least a year and a half. In his early undergraduate notes, he had found many reasons for rejecting Descartes' theory of light as an "ethereal pressure." He tended himself to think of light as made up of material particles. This idea, together with his early discovery that each colour is differently bent or refracted when viewed through a prism, may have been important in leading him to the insight that white light is a mixture of coloured rays. In any event, his discovery followed a much more twisting path than his simplified version suggested. It combined hard, critical examination of existing theories with bold speculations and careful experiments.

Newton's paper created a great stir when the Royal Society read it. But it involved him in quarrels which soured his spirit. He thought he had agreed to share a great discovery with the world only to find himself misunderstood. He had given up "my quiet to run after a shadow."

His critics agreed that to have shown that different colours were differently refracted was a big step forward in optical science. But they questioned his conclusion that the various colours were originally contained in the white light. Robert Hooke said light consisted of a large number of "vibrations," and each vibration produced a different colour when separated out by a prism. The great Dutch follower of Descartes, Christiaan Huygens, was later to put forward a wave-theory much better worked out than Hooke's to give an explanation along these lines.

Replying to his critics, Newton patiently pointed out the ways in which his experimental findings clashed with their different theories. But he was also led to argue as if his own conclusions had resulted purely from experiments. His critics, he said, were prejudiced because they had begun with theories and did not want to give them up. By arguing in this way, Newton was belittling the imagination involved in every great scientific discovery. It made it possible for a misleading view of science, which regards its laws and theories as no more than summaries of un-prejudiced experiments and observations, to appeal to Newton's example.

Christiaan Huygens (1629–1695), the Dutch physicist whose work in dynamics and light ranks him among the greatest of Newton's contemporaries. He is shown here using a tubeless telescope which he designed and built in about 1650.

Pl. 52

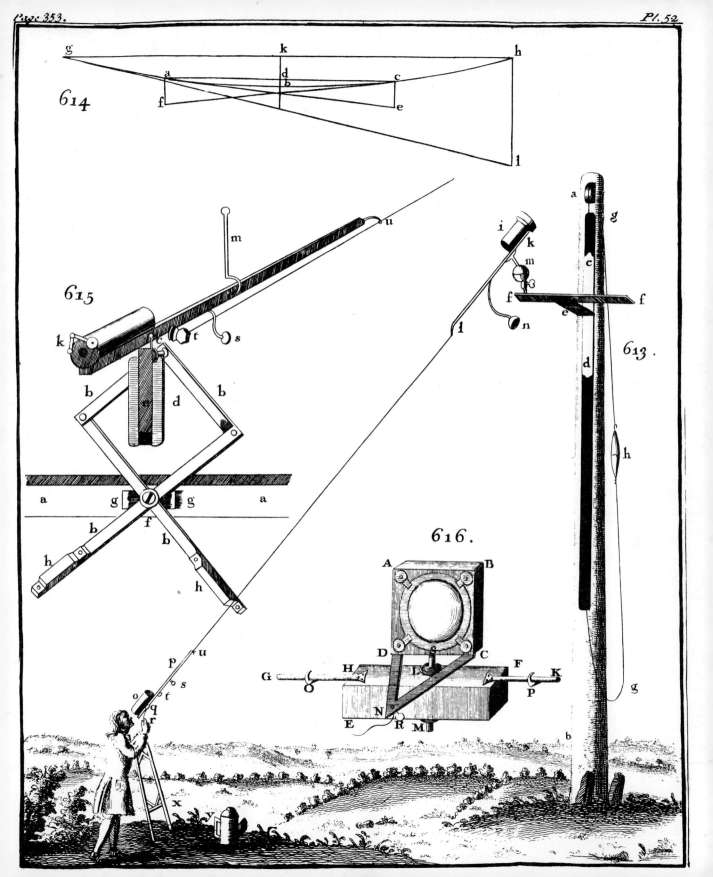

614

615

616.

613.

Newton had not overlooked the possibility of correcting chromatic aberration in lenses by combining different types of glass. But a mistake in an experiment made him think that this would not "disperse" or spread out the spectrum differently. It was almost a hundred years before the mistake was spotted. Dollond managed to make lenses which combined convex crown glass and concave flint glasses. When used as object lenses, they almost cancelled out chromatic aberration. The refracting telescope was again to be the main instrument of astronomers.

Wearied by objections, Newton wrote to Henry Oldenburg that he had considered sending more thoughts on colours to the Royal Society "but find it yet against the grain to put pen to paper any more on that subject." He would "resolutely bid adieu to it eternally . . . " The objections had, however, stung him into ordering his ideas better. He was to make many more discoveries in optics. But they were to be published only in 1704, after the death of his old adversary, Hooke.

Newton's rings which can be seen when two transparent surfaces almost touch — in the pictures below, a flat glass plate and a slightly convex lens. This is one of the many experiments with light which Newton describes and attempts to explain in his *Opticks*.

5. *Gravity and the Laws of Motion*

The story of Newton and the apple has been told and retold. Friends heard it from his own lips in his old age. While sitting in his mother's orchard one day during the plague year, his attention was caught by an apple falling to the ground. This set him thinking, but twenty years were to pass before his ideas about gravity and motion were published in his *Principia Mathematica.*

Why the delay? A number of clever explanations have been given of why he kept his discovery from the world for so long. Recent historical work has changed our ideas about this. It now seems likely that the crucial idea of universal gravitation came to Newton only much later, not until after he had written the first section of his masterpiece.

Newton had already studied the problem of the

Right A descendent of the apple tree under which Newton is said to have sat when an apple fell and set him thinking about gravity.

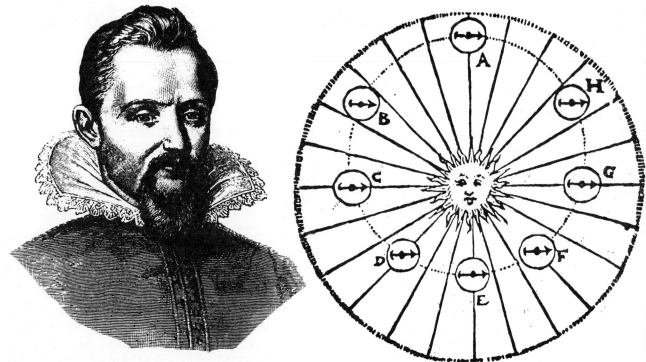

motion of earth and planets around the sun when he was an undergraduate. Aristotle had imagined a nest of hollow spheres carrying sun, moon, and planets around the earth in their daily and yearly motions. Copernicus had not given up the spheres, but his work made others more and more reluctant to believe in them. If there were no spheres what kept the earth and the planets circling around the sun?

Early in the seventeenth century the German astronomer Johannes Kepler suggested that a moving force, radiating from the sun like spokes in a wheel carried the planets round as the sun itself spun round. But he altered this model after a discovery of immense importance. Kepler was inspired by a Pythagoras's belief that mathematics would lay bare the hidden nature of things. He tried to show that the observed positions of the planets were on uniform circular orbits and he tried to do this more exactly than anyone else had done before him. After long

Above left Johannes Kepler (1571–1630), the German astronomer and physicist famous in particular for his laws of planetary motion which provided the basis for much of Newton's work. He thought that the forces (*illustrated above*) which took the planets in fixed orbits round the sun were something akin to magnetism in nature.

Above How to draw an ellipse with a length of cotton and two pins. Each pin will be at a *focus*.

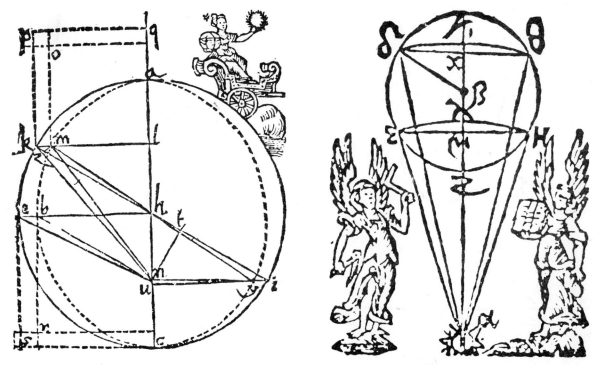

Above Diagrams from Kepler's *Astronomia Nova* (1609) in which, after an immense amount of work, he set out his discovery of the law of elliptical orbits for planets. These particular diagrams explain his discovery of the elliptical orbit of Mars.

The varying velocity in elliptical motion.

years of hard work and failure, he came up with a solution. The ancient idea of uniform circular motion had to be abandoned. The planets moved in *elliptical* orbits (flattened circles) with the sun at one focus. In such an orbit the speeds of the planets and their distance from the sun would change all the time. To explain this Kepler thought of magnetic poles in the planets which were attracted or repelled by a magnetic north pole in the sun itself.

Few people took Kepler's elliptical orbits or his "celestial mechanism" seriously for a long time. A new science of motion had to be created before his ellipses, and two other discoveries, could serve as the foundation of a new world-picture. The first great step towards it was made by another follower of Copernicus, Galileo Galilei.

Galileo is often said to have disproved Aristotle by dropping balls of different weight from the Leaning Tower at Pisa. If Aristotle's theories were right, the

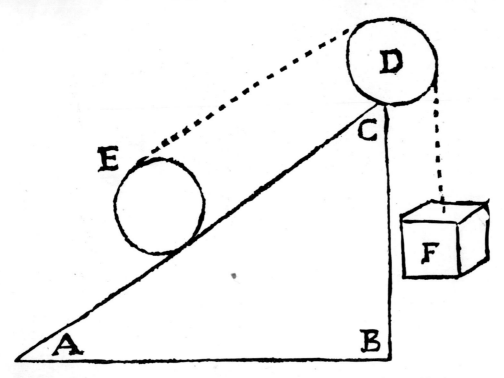

Above Galileo's experiment to measure the force of a ball rolled down a slope. From *Operere di Galileo Galilei* (1665–6).

heavier ball should have struck the ground first. Galileo showed that both balls reached the ground at the same time. The story may or may not be true. It is certainly not a good illustration of Galileo's importance to science. His great achievement was to single out the vital features of physical events in a real-life situation—like a falling stone, or a swinging pendulum—and to "think away" the rest. By doing this he clarified ideas like speed, acceleration and resistance.

Aristotle had stuck close to common experience. He thought that, except for "natural" motions, all movement needed a mover. If the mover stopped pushing the movement would stop. Galileo "thought away" friction and imagined a perfectly smooth ball being placed upon an ideally smooth slope. The ball would roll faster and faster down the slope. On an upward slope, force would be needed to push it along or even to hold it still. Therefore, on an endless level surface, there is no reason why a ball once set moving should ever stop.

Right Galileo Galilei (1564–1650), the Italian astronomer and physicist and one of the major founders of the movement which revolutionized medieval natural philosophy into modern science.

GALILEO GALILEI LINCEO FILOSOFO E MATEMATICO DEL SER.^{mo} GRAN DVCA DI TOSCA.

F. Villamoena Fecit.

59

The principle of *inertia* to which Galileo pointed was a big break with ancient thought. Motion as such does not need a continuous force to keep it going. Only a *change* of motion needs force. Galileo used this principle to answer ancient objections to a moving earth. A stone would not cease to share the motion of the earth the moment it lost contact with it. When dropped from a high tower, it would drop to the foot of the tower because it would really have two motions: one downward, the other a circular one shared with the moving earth. The power of old ideas, even on those trying to break with them, is shown by Galileo's continuing to think of these motions as "natural." Circular motion remained perfect motion for him, only now earthy matter could share this perfection too.

It was Descartes who sharply rejected such ideas. Inertial motion was uniform motion in a straight line. Circular motion was not natural, but needed a mechanical cause. For the earth and the planets, said Descartes, that cause was the whirlpool of subtle matter which carried them around the sun. That same power explained why a stone, when let go, dropped to the ground. The subtle matter in the earth's vortex moved away much faster and the stone, like a piece of wood in a watery whirlpool, was driven to the centre.

If it was the fall of an apple that set Newton thinking in 1666 of the power needed to hold the moon around the earth and the planets around the sun, he probably thought of it as due to these Cartesian vortices. When we whirl a stone in a sling, the tension in the string may seem to us a sign of the stone's tendency to fly away. It came to be called a "centrifugal," or centre-fleeing, tendency. Newton worked out a way of calculating it, and estimated what it was for the earth, moon, and planets. It then occurred to him to combine it with something that Kepler had discovered. Kepler had found that the time (T) taken by a planet to orbit the sun varied and

René Descartes, from his book *Opera Philosophica* (1692).

60

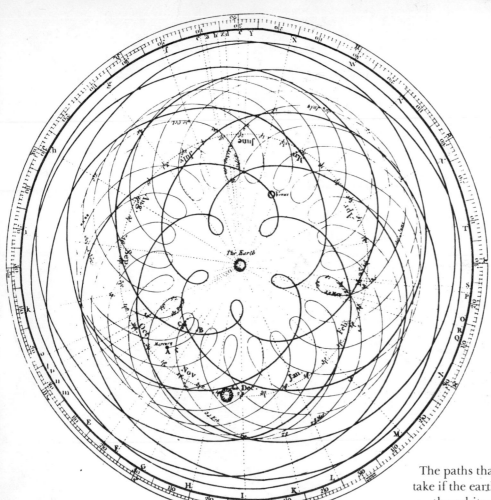

The paths that the planets would seem to take if the earth were stationary. Although the orbits of planets round the sun are nearly circular, seen from any one planet the rest would appear sometimes to move forward, sometimes backward and sometimes to stand still. They would seem to move neither in circles nor in ellipses but in looped curves. This engraving comes from James Ferguson's *ASTRONOMY EXPLAINED UPON Sir ISAAC NEWTON's PRINCIPLES, And made easy to those who have not studied MATHEMATICS,* published in 1764.

depended on its average distance (D) from the sun in such a way that the ratio T^2/D^3 was the same for all planets. That meant, Newton discovered, that the "centrifugal" tendency of the planets must *vary inversely as the square* of their distance from the sun.

It has usually been assumed that Newton must have thought of the centrifugal tendency as balanced by a gravitational force which acts from the sun upon the earth and the planets. As that force is universal, it should be possible to test the effect of the earth's gravitational force upon the moon. The moon's inertial motion would carry it with uniform speed in a straight line. It orbits the earth because it is con-

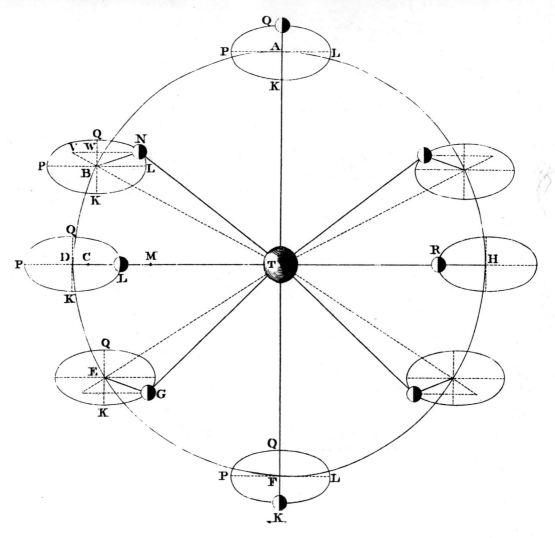

One of three diagrams from a postscript by John Machin to an 1803 edition of Newton's *Principia* illustrating "The Law of the Moon's Motion According to Gravity.

tinuously pulled away from that path by the earth's gravitation. Newton knew that at the surface of the earth that gravitational pull causes objects, such as apples, to fall sixteen feet in the first second after starting from rest. If the pull decreases according to the inverse-square law, then at the distance of the moon its force would be 1/3600 of that on the earth. An object on the moon must then take one minute to fall the distance it fell in one second on the earth. So according to gravity, the moon should be pulled away from its straight inertial path by sixteen feet every minute. Newton tested the result by a simple calculation. He worked out the moon's acceleration, from its period

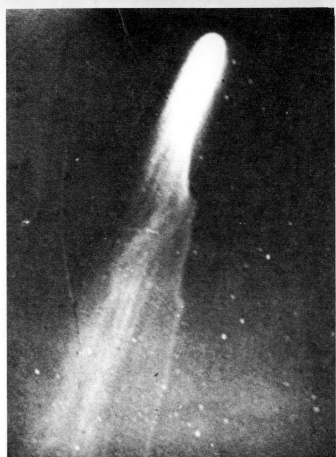

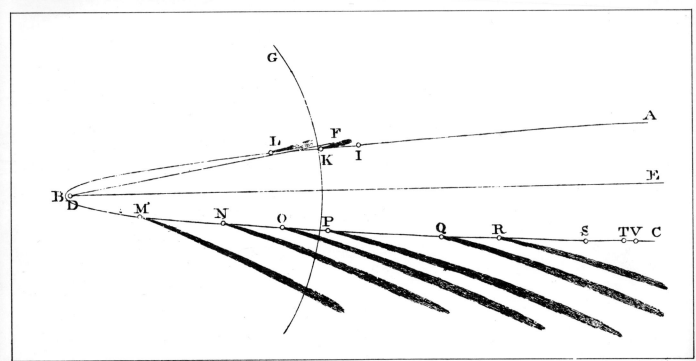

Top left Edmond Halley (1656–1742) from the portrait by Michael Dahl, painted in 1736 when Halley was eighty. Halley's great contribution to astronomy was in the study of comets. He observed the great comet of 1680 in Paris and worked out its orbit. After noting the similarity of the orbits of the comets of 1456, 1531, 1607 and 1682, he concluded that they were one and the same comet and predicted to within five months its return at the end of 1758.

Top right is a photograph of Halley's comet when it returned yet again in 1910.

Below is Newton's diagram of the orbit of the 1680 comet.

of revolution and the supposed size of its orbit around the earth. He then compared the acceleration worked out in this way, with that given by the inverse-square relation. The answer agreed "pretty nearly", but not exactly, since (as we now know) he used too small a figure for the size of the earth. Recent historical work has, indeed, unearthed a moon-test by Newton, probably done in 1666, but there is no suggestion there that he had begun to think of the inward force as a gravitational attraction between sun and planets, or earth and moon.

Newton put aside work on these problems, until brought back to them in 1679 by Robert Hooke, who had now become Secretary of the Royal Society. Hooke asked him to suggest problems to revive public interest in science. Hooke wrote to Newton suggesting that if a body orbited the earth so that it was *attracted* by the earth with a force varying inversely as the square of the distance, its path would be an ellipse with the earth at one focus. Hooke offered no mathematical proof. Newton later claimed that he had worked out a proof but had tossed the paper aside, "being upon other studies."

In 1684 Edmond Halley visited Newton in Cambridge. Halley had found that Hooke and Sir Christopher Wren, like himself, had guessed that Kepler's sun-planet relation meant that there is an inverse-square attraction between sun and planets. None of them could offer a mathematical proof. Hooke claimed one, but did not take up Wren's wager of a book worth forty shillings.

Halley was delighted when Newton told him not only that he had made the same discovery, but that he had proved it mathematically for elliptical orbits. He could not find the proof among his papers, but promised to reconstruct it. The result was a tract which Halley saw on a second visit some months later. Newton's mathematical genius was displayed in his solution of a problem that had baffled the others. But the crucial idea of universal gravitation, which appears in the *Principia*, was still absent from this tract. To appreciate his final achievement we must grasp the problems which faced him.

Robert Hooke saw that the principle of inertia meant that a force must be acting whenever a body moves in any way other than with uniform motion in a straight line. For example, circular motion needs a force to act continuously towards the centre of the circle. Instead of concentrating upon supposed "centrifugal" forces arising from orbital motion, Hooke pointed to the force that made a body travel in such a path. For the earth and the planets, Hooke suggested that this force was an attraction towards the centre of the sun.

But a deflecting force of that sort was unacceptable to "mechanical philosophers." Galileo had rejected Kepler's ideas of an attractive force from the sun. He had even denied the old idea that the attraction of the moon causes tides on earth and he stubbornly defended his own theory that tides were caused by the earth's rotation. Descartes criticized Roberval for explaining the fall of heavy bodies to earth by attraction. He could not conceive of a force acting at a distance as gravitational force was supposed to do: there must be a mechanical explanation. Such ideas seemed to Galileo and Descartes to belong to the magical tendencies which had gripped the Renaissance imagination as the hold of Aristotle slackened.

Newton himself had up to now been faithful to Descartes' programme of explaining all that happened in nature by matter in motion. His acceptance of the idea of gravitational attraction plunged him into controversies to the end of his life. He could not decide whether to explain gravity as due to the direct action of God, or as due ultimately to matter in motion. What first caused him to consider such a departure from the mechanical programme?

One reason may have been the chemical and

Isaac Newton a year before his death by J. Vanderbank.

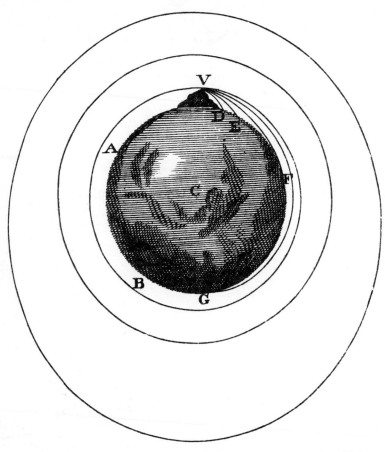

alchemical studies on which he spent so much of his time and which have continued to puzzle historians. Newton was deeply impressed by the way some chemical substances seem to have an attraction for certain others and readily combine with them. Certain other things were difficult to explain by mechanisms of the Cartesian sort. Why does matter stick together and not just fall apart? Why does water rise in thin glass tubes? Why are light rays bent away from their paths when they enter another substance, and how can flies walk on water without wetting their feet? Attractions and repulsions between particles of matter could explain all this more simply.

Was that a backward step? Not, Newton thought, if we could discover the small number of mathematical laws presumably ruling such attractions and re-

Left A diagram from Newton's *System of the World* showing the paths of projectiles sent off at various velocities until a high enough one is reached to put the particle in orbit.

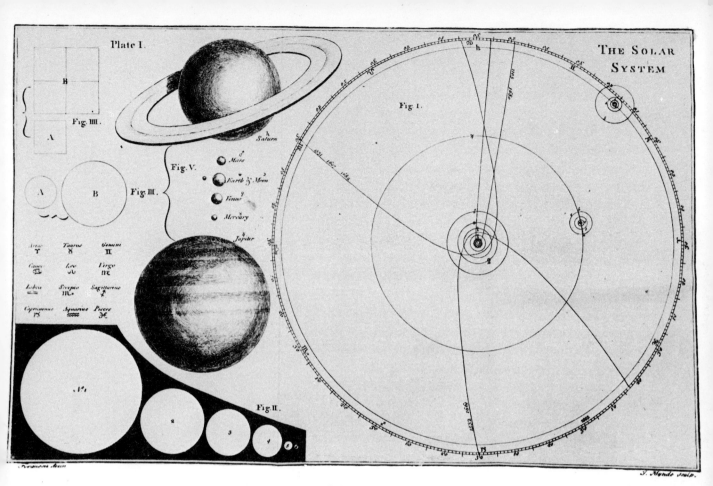

Above The solar system as it appeared in 1764, including the relative sizes of the known planets. From Ferguson's *Astronomy Explained . . .* (1764).

pulsions. Matter and motion are *not* enough to explain nature. In addition, there must be certain "active principles" planted by God in nature. Without their continuing action nature would have come to a standstill long ago. Newton's scientific and religious thought came together in this view of nature which he was developing. He was deeply influenced by the thinkers known as the Cambridge Platonists. They believed that if the universe was thought of as a clockwork which God had created but then left to run itself by mechanical principles, then men would stop believing in religion. His thoughts already seem to have been moving away from purely Cartesian explanations of nature just at the time when he began once again to think about the movements of the heavens.

If the sun pulls on the planets with a gravitational

force, is this a force that acts between *particular* bodies (like the specific attractions in the chemical reactions Newton had studied)? Or did *all* matter attract *all other* matter? Newton decided that the force was universal.

How was the pull related to the bigness or smallness of bodies? In answering that, Newton made the first clear distinction between *weight* and *mass*. The weight of a body measures the gravitational pull, which will vary with distance from the centre of the earth. The mass is the "quantity of matter" in a body. A body may weigh less at the top of a high mountain, but its mass need not have changed.

Mass and weight must be exactly proportional at a fixed distance from the centre of the earth. Kepler's law connecting the distances of the planets from the sun with the time they took to orbit it meant that if all the planets were placed at the same distance from the sun, they would move in the same orbit. Newton realized that this would happen only if the pull of the sun was exactly proportional to their very different masses. It would pull, say, four times as hard on a body four times as massive as another at the same distance.

Galileo had scandalized Aristotelians by saying that, if it were not for air resistance, all bodies falling from the same height would have the same speeds. Newton now showed that to achieve this the gravitational force would have to pull harder on a more massive body, since there was more matter to be moved. Had the strength of the pull been the same, the *lighter* body would have fallen faster. It was a surprising fact that the pull was, in fact, exactly proportional to the mass of bodies. Newton further tested the proportionality by experimenting with pendulums, whose usefulness in dealing with problems of falling bodies had already been seen by Galileo. Galileo had also shown that the speed of a falling body increases regularly with time (uniform acceleration). This could now be shown to illustrate the general principle that a

The type of pendulum used by Newton to investigate how mass is related to gravitational pull. The engraving is from the second edition of W. J.'s Gravesande's *Physices Elementa Mathematica* (1725).

TAB. XVII

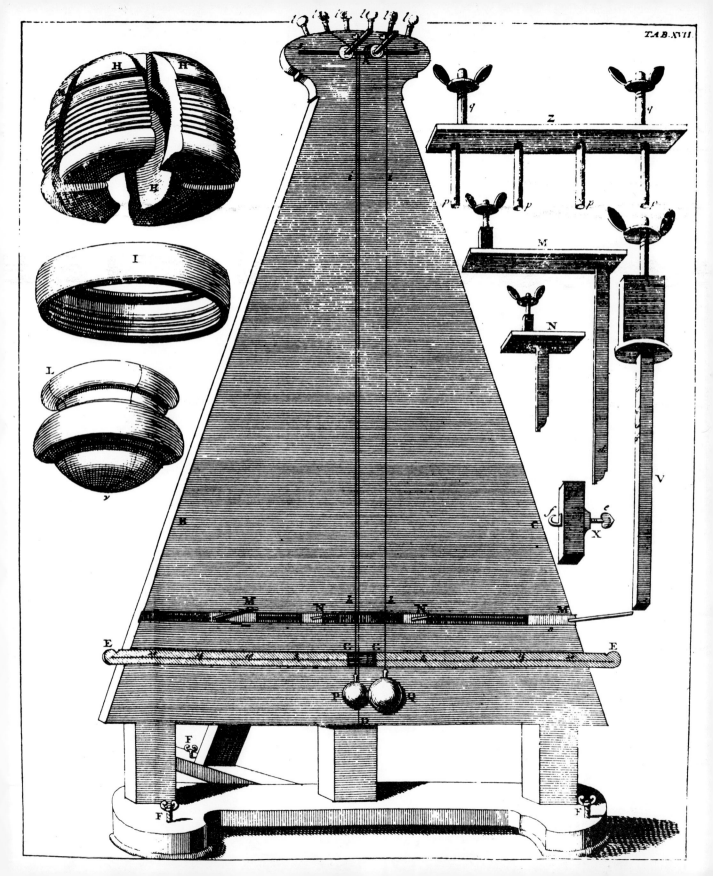

constant force produces, not a constant velocity as in the old physics, but a constant *increase* in velocity.

When calculating forces between massive bodies at great distances, like the sun and the planets, or the earth and the moon, it was possible to treat them as "mass-points." But calculating the forces upon, say, an apple falling to the earth, seemed to be very complicated. Parts of the earth near at hand would pull hard, while other parts would pull with a force that became weaker at a distance. Using his new mathematical tool of fluxions, Newton proved that such massive and solid spheres attracted things *as if* their mass was concentrated at their centres. In this way the problems involved were enormously simplified.

Gravitation was a "centripetal" force. A more general idea of force was needed to create a workable science of dynamics, that is a science of motion and moving things. That was a very tough problem which had defeated earlier pioneers. There were so many different ideas of forces, measured in very different ways. Newton's true genius comes across in the simplicity of his solution. All matter possessed "inertia"—the power of resisting motion when at rest, or changes (of direction and magnitude) when in motion. Gravitation was an example of an "impressed" force which changed inertial motion. Forces, said Newton must be proportional to, and can be measured by, the changes they cause in such inertial motion.

Newton set out the basic laws of his new mechanics in three laws of motion. The first stated that every body continues in its state of rest or of uniform motion in a straight line unless made to change that state by forces impressed upon it. The second law related the increase in velocity of a body to the impressed force: the acceleration gained by a body is proportional to the force and inversely proportional to its mass. The third law stated that to every action there is an equal and opposite reaction.

Left Isaac Newton's rooms at Trinity College, Cambridge. *Above* Roubillac's statue of Newton in Trinity College.

Newton now applied his new ideas to a great range of problems. Besides free fall and impact, he analysed the much harder problems of resistance, wave motion, and the motion of fluids. He tackled the "three-body" problem, that is the problem of orbital motions where more than two bodies are involved. The motions of the earth and the planets were obviously of this sort. He was able to explain the tides, as well as the long-term motion of the axis of the earth which produced a "precession of the equinoxes" in the heavens, and predicted a bulge at the earth's equator.

Newton struck the death-blow to Cartesian vortices in the course of his investigations. He worked out that they would lose motion continuously and could not carry on for very long. Nor could any Cartesian "subtle matter" explain gravitation. The presence of an atmosphere even as thin as that of the earth in interplanetary space would make Jupiter lose one-tenth of its motion in thirty days. Instead of the full universe of Descartes, Newton's was an almost empty universe, where the emptiness throughout space and within physical objects quite overshadowed the small parcels of matter scattered through its vast reaches.

These mammoth achievements were the result of just two years' work. By April 1686 the manuscript of the *Mathematical Principles of Natural Philosophy* was before the Royal Society. Halley decided to print it at his own expense when the Society found itself unable to pay for it. Besides his financial sacrifice, Halley also had to soothe Robert Hooke, who now accused Newton of stealing the inverse-square law from him. "Philosophy," Newton sadly commented, "is such an impertinently litigious Lady, that a man had as good be engaged in lawsuits, as to have to do with her." Halley had a hard time persuading Newton not to leave out the section *System of the World* which was the crown of his work.

The great work was published in May 1687 and

The earth's axis of rotation (as shown by the dotted line) slowly changes in direction with respect to the stars. Each full turn takes about 26,000 years.

Above Gravity re-discovered. A *Punch* cartoonist's view.

PHILOSOPHIÆ
NATURALIS
PRINCIPIA
MATHEMATICA.

Autore *J*S. *NEWTON* Trin. Coll. Cantab. Soc. Matheseos Professore *Lucasiano*, & Societatis Regalis Sodali.

IMPRIMATUR·
S. PEPYS, *Reg. Soc.* PRÆSES.
Julii 5. 1686.

LONDINI,

Jussu *Societatis Regiæ* ac Typis *Josephi Streater*. Prostat apud plures Bibliopolas. *Anno* MDCLXXXVII.

Above The title page of Newton's *Principia,* published in 1687. Edmond Halley, whose constant encouragement Newton had received while preparing the manuscript, paid for its publication out of his own pocket when the Royal Society found itself unable to pay for it.

created an immediate sensation. Even the followers of Descartes, who rejected its central idea of an attractive gravitational force, came in time to recognize its achievement. The dream of a mathematical science of nature was at last set on firm foundations, even if at its heart lay a type of action which the founders of the mechanical philosophy had wished to banish from any rational system of science.

A children's version of Newtonian physics
published in 1761 by the Lilliputian Society
and measuring about 3 × 5 inches.

Frontifpeice.

Lecture on Matter & Motion.

THE
NEWTONIAN SYSTEM
OF
PHILOSOPHY

Adapted to the Capacities of young
GENTLEMEN and LADIES, and fami-
liarized and made entertaining by Objects with which they are intimately
acquainted:

BEING
The Substance of SIX LECTURES read to the
LILLIPUTIAN SOCIETY,

By TOM TELESCOPE, A.M.

And collected and methodized for the Benefit
of the Youth of these Kingdoms,

By their old Friend Mr. NEWBERY, in *St.
Paul's Church Yard*;

Who has also added Variety of Copper-Plate Cuts, to
illustrate and confirm the Doctrines advanced.

---

*O Lord, how manifold are thy Works! In Wisdom
hast thou made them all, the Earth is full of thy
Riches.*

*Young Men and Maidens, Old Men and Children,
praise the Lord.*   PSALMS.

LONDON,
Printed for J. NEWBERY, at the BIBLE and SUN,
in *St. Paul's Church Yard.* 1761.

6. *A Very Eminent Citizen*

Newton was forty-four years old when his *Principia* was published. His assistant and secretary, Humphrey Newton, remembered the intense concentration that went into writing it. Newton seldom slept before two or three o'clock in the morning, and, when doing chemical experiments, not till five or six. He would start his work again after resting for only four or five hours. He would pace his room so restlessly that "you might have thought him to be educated at Athens among the Aristotelean sect" (it was Aristotle's custom to walk up and down while teaching at the Lyceum). He often left untouched what food was prepared for him. On the rare occasions when he dined in the college hall, if not reminded he "would go very carelessly, with shoes down of heels, and his head scarcely combed . . ."

Few people went to the small number of lectures Newton gave as Lucasian Professor. Sometimes he came back from an empty hall. Early in 1687 his work was interrupted when he went to London with the Vice-Chancellor and others from Cambridge University. James II, a convert to Catholicism, had been on the throne for two years. Cambridge refused his order to give a degree to a Benedictine monk, and its officials were summoned to appear before a notorious judge, George Jeffreys. Jeffreys would not listen to their explanations, and the Vice-Chancellor was deprived of his offices. In November 1688 William of Orange landed in England and King James was forced into exile. The Vice-Chancellor was restored, and Newton served as the Member for Cambridge in the Parliament which sat from 1689 for little over a year.

The work of composing the *Principia,* together with

Left George Jeffreys (1648–89), the notorious judge who was President of the High Court of Commission and although not a Catholic himself, he proved himself a willing tool of James II's plan to force Britain round to the Roman Catholic viewpoint. The engraving here shows Jeffreys in 1688 when, following the King's example he tried to flee the country but was caught at Wapping disguised as a sailor.

pressures in his personal and intellectual life, led to Newton's mental breakdown in 1693. He is said to have spoken only once during his year in Parliament, and then only to ask an usher to open a window. But he had influential friends, especially Charles Montague, a former student, whose political star rose rapidly. His friends tried to get Newton a public post, at first without much success. During the same year Newton went back to Grantham to nurse his mother devotedly during her last illness. Newton had again taken up his theological work, but tried to suppress a work challenging the orthodox view of the Holy Trinity which he had earlier been anxious to publish in Holland in translation without putting his name to it.

In 1693 his friends Samuel Pepys (famous for his diary) and the philosopher John Locke were startled to receive strange letters from Newton. Pepys wrote at

Above Samuel Pepys (1633–1703), the famous diarist, a founding member of the Royal Society and a close friend of Newton.

once to Cambridge to ask a friend to visit Newton. The letter had greatly disturbed him, "lest it should arise from that which of all mankind I should least dread from him and most lament for . . ." But Newton's recovery was rapid and his mental powers, soon tested in scientific controversies and new public duties, showed their old vigour.

Newton said goodbye to Cambridge in 1695 to take up the Wardenship of the Mint which Montague, now Chancellor of the Exchequer, managed to get for him. The post was no mere sinecure. Montague had decided to reissue all the coin of the realm. Widespread clipping of gold and silver coin had devalued English currency and become a scandal. Clipped coin was now called in, melted, and reissued with milled edges at the Mint in London and at branches in several towns. Only firstclass management made it possible to do the job in two years, in the midst of a war with France and with continuing hostility from the Parliamentary Opposition.

Below The minting of coins. Newton was Warden of the Royal Mint from 1695 and was promoted to Master of the Mint in 1699.

SIR,

THESE are to give Notice, That on *Monday* the First Day of *December* 1712, (being the next after St. *ANDREW's DAY*) the Council and Officers of the ROYAL SOCIETY are to be Elected for the Year ensuing; at which ELECTION your Presence is expected, at Nine of the Clock in the Forenoon, at the House of the ROYAL SOCIETY, in *Crane Court,* *Fleet Street.*

To
Thomas Jffed Efq;

Is. Newton P.R.S.

Although Montague's party was defeated at the next election, the Tories kept Newton on and made him Master of the Mint in 1699.

Newton had lost his Parliamentary seat during the election of 1690. He was reelected for Cambridge in 1701. Two years later he received the honour of being elected President of the Royal Society. He was reelected President every year for the remaining twenty-five years of his life.

Above A Royal Society election notice for December 1712, signed by the President of the Society, Isaac Newton.

Right A meeting of the Royal Society in Crane Court, Fleet Street, presided over by Newton.

OPTICKS:

OR, A
TREATISE
OF THE
REFLEXIONS, REFRACTIONS,
INFLEXIONS and COLOURS
OF
LIGHT.

ALSO
Two TREATISES
OF THE
SPECIES and MAGNITUDE
OF
Curvilinear Figures.

LONDON,

Printed for SAM. SMITH. and BENJ. WALFORD.
Printers to the Royal Society, at the *Prince's Arms* in
St. *Paul's* Church-yard. MDCCIV.

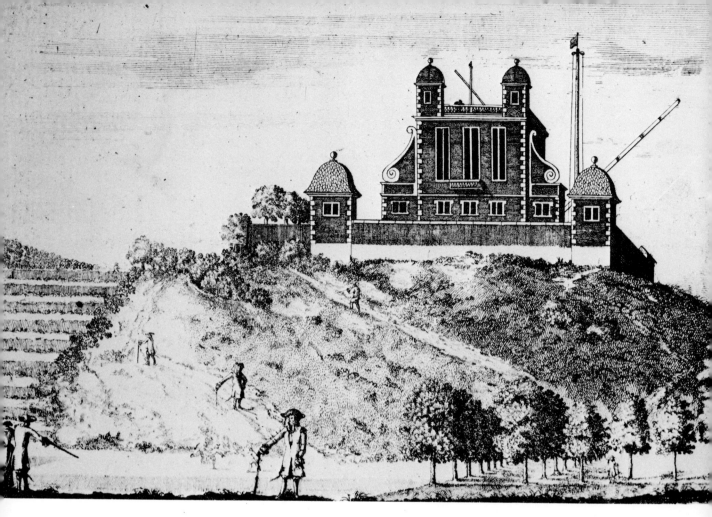

Above Flamsteed House, built in 1676 as the Royal Observatory for John Flamsteed, the first Astronomer Royal.

Left The title page of Newton's great popular work on light, *Opticks,* published in 1704.

While busy with the recoinage, Newton had written to the Astronomer Royal, John Flamsteed, criticizing him for spreading the rumour that he, Newton, would soon offer an improved theory of the motion of the moon. He did not, he said, wish to be "teased by foreigners about mathematical things, or to be thought by our own people to be trifling away my time about them, when I should be about the King's business." But he had not turned away from science. In 1704 his *Opticks* was published. Contrasting with the highly technical and mathematical *Principia,* it long remained a popular work. Newton worked hard upon revisions for the second and third editions of the *Principia* which appeared in his lifetime, edited by younger mathematicians.

At Cambridge his assistant had heard Newton laugh only once. That was when someone asked him what was the use of studying Euclid. He became more friendly and cheerful in London society, "easily made to smile, if not to laugh." His encouragement of the talents of younger friends was one of his most attractive features. But there was plainly another side to his character.

John Locke, who greatly admired him, described Newton to a friend as "a little too apt to raise in himself suspicions where there is no ground ..." Flamsteed, a less impartial witness, called him "insidious, ambitious, and excessively covetous of praise, and impatient." Flamsteed had reason for his bitterness. When he angered Newton by disagreements, Newton had seen to it that Flamsteed would have no control over the printing of his life's work in astronomy. The great German philosopher

Above left John Flamsteed (1646–1719) and *above* Gottfried Wilhelm von Leibniz (1646–1716), two men who were targets of Newton's anger. Flamsteed, the Astronomer Royal had annoyed Newton with his disagreements to such an extent that Newton took out of Flamsteed's control the publication of his life's work in astronomy. Leibniz was Newton's great rival to the claim to the discovery of calculus.

The inauguration ceremony of Newton's statue in Grantham, Lincolnshire, in 1858.

and man of science, Leibniz, was another man who became the target of Newton's anger. Newton had once admitted that each of them had discovered the calculus on their own at about the same time. Once a dispute on priority had been fanned by others, Newton masterminded a supposedly neutral committee which the Royal Society appointed at Leibniz's request. Leibniz's conduct in the affair was not blameless. But he was made to seem to have stolen his discovery from Newton.

Carried in his extreme old age by coach or sedan chair to preside over meetings of the Royal Society, Newton outlived old enemies like Hooke, Flamsteed, and Leibniz. Even in the country of Descartes, leading younger philosophers like Voltaire wished to sweep the Cartesian vortices from the scientific skies to leave only the gravitational attraction and empty space of Newton. Locke and Newton became patron saints of the great movement of thought known as the Enlightenment. When the Romantic Movement rejected the ideals of that "Age of Reason" in the later eighteenth century, Newton inevitably became for them a symbol of a narrow and cold rationality. For the Romantics, deep feeling and a sense of mystery in life were more important in understanding the world than a mathematical analysis of its motions.

Neither Enlightenment nor Romantic image did full justice to the Newton whom the poet William Wordsworth imagined voyaging over "strange seas of thought." Like many other great innovators, he turned his face to the past even as he pointed the way forward. His religious roots lay in Protestant fundamentalism, which means he believed everything in the Bible to be true beyond doubt. The Biblical studies on which he worked for many years aimed to prove that the prophetical books had foretold the course of ancient history in the most accurate detail. While eager in the defence of his scientific claims, he seriously thought about putting forward evidence in the second edition of the *Principia* to prove that the law of universal gravitation was hidden in the ancient idea of the "music of the spheres," and that Pythagoras had solved that riddle. He felt at home in the world of the alchemist, and clung to a world of ideas which his own work was to do most to destroy, and to which the Romantics would seek to return.

Newton suffered ill health from about 1722. In 1725 he moved to Kensington from his home in Jermyn Street, Piccadilly, to breathe a purer country air. In February of that year he felt well enough to preside over a meeting of the Royal Society in town. His last illness soon followed. He sank into unconsciousness on 18th March and died early in the morning of the 20th, being then in his eighty-fifth year. He was buried after a magnificent funeral in Westminster Abbey. "Let mortals rejoice," read the words upon the monument erected to him in 1731, "that there has

OBSERVATIONS
UPON THE
PROPHECIES
OF
DANIEL,
AND THE
APOCALYPSE
OF
St. JOHN.

In Two Parts.

By Sir *ISAAC NEWTON.*

LONDON,
Printed by J. DARBY and T. BROWNE in *Bartholomew-Close.*
And Sold by J. ROBERTS in *Warwick-lane*, J. TONSON in the *Strand*, W. INNYS and R. MANBY at the West End of St. *Paul's Church-Yard*, J. OSBORN and T. LONGMAN in *Pater-Noster-Row*, J. NOON near *Mercers Chapel* in *Cheapside*, T. HATCHETT at the *Royal Exchange*, S. HARDING in St. *Martin's lane*, J. STAGG in *Westminster-Hall*, J. PARKER in *Pall-mall*, and J. BRINDLEY in *New Bond-street.*
M,DCC.XXXIII.

Above The title page to one of Newton's works on religion.

Right Newton's house in Kensington into which he moved in 1725.

existed such and so great an ornament of the human race.''

Perfected by the work of eighteenth- and nineteenth-century scientists, Newton's system of the world appeared to be his permanent monument, until Albert Einstein, in our own century, made it a special case of a more comprehensive theory. Newton himself searched for a knowledge that went beyond any physical theory.

''I do not know,'' said Newton, ''what I may appear to the world; but to myself I seem to have been only like a boy, playing on the seashore, and diverting myself in now and then finding a smoother pebble or a prettier shell than ordinary, while the great ocean of truth lay all undiscovered before me.''

Newton's death mask.

Date Chart

1642	Born on Christmas Day at Woolsthorpe, Lincolnshire.
1654	Admitted to King's School, Grantham, and lodged with apothecary in town.
1656	Recalled from school on death of step-father and return of mother to Woolsthorpe.
1660	Prepares for Cambridge at Grantham.
1661	Admitted sub-sizar at Trinity College, Cambridge.
1665	Bachelor of Arts. Plague causes return to Woolsthorpe except for perhaps short visit in 1666.
1667	Return to Cambridge. Master of Arts.
1668	Elected Senior Fellow of Trinity College.
1669– 1671	Appointed Lucasian Professor of Mathematics to succeed Barrow.
1671	Sends reflecting telescope to Royal Society.
1672	Sends new theory of light and colours to Royal Society. Elected Fellow of Royal Society.
1674	Permitted by Royal patent to hold College Fellowship without being ordained, while Lucasian Professor.
1676	Binomial Theorem sent to Royal Society.
1679	Letter from Hooke revives Newton's interest in mechanics.
1684	Halley's visit to Newton and promise of proof of inverse-square law for elliptical orbits.
1686	Printing of Book I of *Principia Mathematica Philosophia Naturalis* begun at Halley's expense.
1687	Books II and III reach the Royal Society in

	March and April, *Principia* appears in July.
→1696	Appointed Warden of Royal Mint in London. Engaged in recoinage.
→1699	Master of Mint.
1701	Finally resigns Lucasian Professorship.
→1703	President of Royal Society.
1704	*Opticks* published.
→1705	Knighted by Queen Anne at Cambridge.
1713	Royal Society committee reports on calculus priority dispute between Newton and Leibniz. Second edition of *Principia*.
1724	Moves from Piccadilly to village of Kensington.
1726	Third edition of *Principia* published.
1727	Newton dies and is buried in Westminster Abbey.

Glossary

ABERRATION, CHROMATIC AND SPHERICAL The failure of an optical system to make the rays of light come to one focus. It is due to the varying thickness of the lens in *spherical* aberration; in *chromatic* aberration bacause the different colours that make up white light are differently bent as they pass through the lens. The resulting image will be fuzzy.

ALCHEMY The art and science which aimed at the ''perfection'' of matter through chemical operations understood in religious terms. The alchemist who could turn a base metal into gold was also supposed to have attained religious self-perfection.

APOTHECARY The ancestor of our druggist or pharmaceutical chemist. He prepared and sold drugs for medicinal purposes.

BAROMETER An instrument for measuring the weight or pressure of the atmosphere. It may be used to measure altitude, or, more commonly, to help in forecasting the weather.

CENTRIPETAL AND CENTRIFUGAL FORCE To make a body move in a circle, it must be pushed or pulled inward by agents like a string, a spring, or gravity. Such a force is called a *centripetal* or centre-tending force. When a stone is whirled on a string, the string seems to pull the hand outward: that is a real *centrifugal* or centre-fleeing force on the hand, although *not* on the stone, as the only force acting on it is inward. It is equal and opposite to the centripetal force of the stone.

CONCAVE A *concave* surface is hollow, or like the inside of a circle or sphere.

CONVEX A *convex* surface is its reverse, and is a curve that bulges out.

COSMOLOGY   A *cosmology* is a system of the universe and of the general laws which govern it.

ELLIPTICAL   An *elliptical* orbit has the shape of an ellipse, that is, a closed curve in which the sum of the distances of any point from the two foci is a constant quantity.

ETHER   For Aristotle a fifth element, a superior to the four elements making up everything below the sphere of the moon. All stars, planets and other heavenly bodies were made up of the Ether.

GRAVITATION   Newton's law of universal gravitation states that every particle of matter in the universe attracts every other with a force that varies directly as their masses and inversely as the square of the distance between them.

INERTIA   The property of a body by which it tends to persist in a state of rest or of uniform motion in a straight line.

INFINITESIMAL CALCULUS The mathematical method used to calculate the rate of change of continuously varying quantities by treating the infinitesimal differences between consecutive values of such quantities.

LAWS OF MOTION   The basic laws of the science of mechanics, like Descartes' laws of impact and pressure, or the three laws of motion given by Newton in his *Principia Mathematica*.

MASS The quantity of matter in a body, as distinguished from its weight. It is the quantitative measure of inertia.

MECHANICS The branch of applied mathematics which deals with the motions of bodies. It includes the study of the forces that cause motion, or *dynamics,* and the study of motion without reference to the forces causing it, or *kinematics*.

REFRACTION The change of direction of a ray of light as it passes from one transparent medium to another, as from air to water. *Refrangible* means being capable of being refracted.

THE SINE LAW OF REFRACTION When a ray of light
strikes obliquely the surface which separates two
transparent media, it is bent in such a way that the
sine of the angle of incidence to the sine of the angle
of refraction is always constant for those two
media.

SPECTRUM The coloured band into which a beam of
light is decomposed by means of a prism.

THEOLOGY "The science of things divine," dealing
with God, His nature and attributes, and His
relations with man and the universe.

WEIGHT The weight of a body measures the force
exerted by gravity on that body.

Further Reading

The standard biographies are:

Sir David Brewster, *Memoirs of the Life, Writings, and
Discoveries of Sir Isaac Newton,* Edinburgh, 1855, two
volumes.

L. T. Moore, *Isaac Newton, A Biography,* London, 1934,
Dover ed. 1962.

A lively but more controversial account is:

F. E. Manuel, *A Portrait of Isaac Newton,* Cambridge,
Mass., 1968.

Newton's work is discussed in:

J. Herivel, *The Background to Newton's Principia,*
Oxford, 1965.

A. Koyré, *Newtonian Studies,* London, 1965.

F. E. Manuel, *Isaac Newton Historian,* Cambridge,
1963.

R. S. Westfall, *Force in Newton's Physics,* London, 1971.

D. T. Whiteside, (ed.), *The Mathematical
Papers of Isaac Newton,* Vols I– , Cambridge,
1967– [in progress].

Picture Credits

The author and publisher wish to thank all those who
have given permission for their illustrations to appear
n the following pages: Grantham Public Library:
, 12, 12–13, 13, 15, 16, 34, 40, 62, 64, 67, 69, 75, 76,
, 86; Mary Evans Picture Library: 78, 84; *Punch*: 74;
dio Times Hulton Picture Library: 80; Ronan
ture Library: 29, 36–37, 43, 46–47, 48, 50, 55, 56,
61, 64, 71, 79, 84; The Science Museum: 44–45,
54, 83; C. V. Wroth: 14, 33, 35, 39, 56, 57, 81,
The remaining pictures belong to the Wayland
ure Library.

Picture Credits

The author and publisher wish to thank all those who have given permission for their illustrations to appear on the following pages: Grantham Public Library: 11, 12, 12–13, 13, 15, 16, 34, 40, 62, 64, 67, 69, 75, 76, 85, 86; Mary Evans Picture Library: 78, 84; *Punch*: 74; Radio Times Hulton Picture Library: 80; Ronan Picture Library: 29, 36–37, 43, 46–47, 48, 50, 55, 56, 58, 61, 64, 71, 79, 84; The Science Museum: 44–45, 53, 54, 83; C. V. Wroth: 14, 33, 35, 39, 56, 57, 81, 87. The remaining pictures belong to the Wayland Picture Library.

Index

THE SINE LAW OF REFRACTION When a ray of light strikes obliquely the surface which separates two transparent media, it is bent in such a way that the sine of the angle of incidence to the sine of the angle of refraction is always constant for those two media.

SPECTRUM The coloured band into which a beam of light is decomposed by means of a prism.

THEOLOGY "The science of things divine," dealing with God, His nature and attributes, and His relations with man and the universe.

WEIGHT The weight of a body measures the force exerted by gravity on that body.

Further Reading

The standard biographies are:

Sir David Brewster, *Memoirs of the Life, Writings, and Discoveries of Sir Isaac Newton,* Edinburgh, 1855, two volumes.

L. T. Moore, *Isaac Newton, A Biography,* London, 1934, Dover ed. 1962.

A lively but more controversial account is:

F. E. Manuel, *A Portrait of Isaac Newton,* Cambridge, Mass., 1968.

Newton's work is discussed in:

J. Herivel, *The Background to Newton's Principia,* Oxford, 1965.

A. Koyré, *Newtonian Studies,* London, 1965.

F. E. Manuel, *Isaac Newton Historian,* Cambridge, 1963.

R. S. Westfall, *Force in Newton's Physics,* London, 1971.

D. T. Whiteside, (ed.), *The Mathematical Papers of Isaac Newton,* Vols I– , Cambridge, 1967– [in progress].